JOHN GLOAG ON INDUSTRIAL DESIGN

Volume 3

THE MISSING TECHNICIAN IN INDUSTRIAL PRODUCTION

THE MISSING
TECHNICIAN
IN INDUSTRIAL
PRODUCTION

JOHN GLOAG

Routledge
Taylor & Francis Group

LONDON AND NEW YORK

First published in 1944 by George Allen & Unwin Ltd.

This edition first published in 2023
by Routledge
4 Park Square, Milton Park, Abingdon, Oxon OX14 4RN

and by Routledge
605 Third Avenue, New York, NY 10158

*Routledge is an imprint of the Taylor & Francis Group, an informa
business*

© 1944 John Gloag

British Library Cataloguing in Publication Data
A catalogue record for this book is available from the British
Library

ISBN: 978-1-032-36309-7 (Set)
ISBN: 978-1-032-36588-6 (Volume 3) (hbk)
ISBN: 978-1-032-36589-3 (Volume 3) (pbk)
ISBN: 978-1-003-33278-7 (Volume 3) (ebk)

DOI: 10.1201/9781003332787

Publisher's Note
The publisher has gone to great lengths to ensure the quality of this
reprint but points out that some imperfections in the original copies
may be apparent.

Disclaimer
The publisher has made every effort to trace copyright holders and
would welcome correspondence from those they have been unable
to trace.

THE MISSING
TECHNICIAN

IN INDUSTRIAL PRODUCTION

By
JOHN GLOAG

With an Introduction by
CHARLES TENNYSON
C.M.G.

LONDON
GEORGE ALLEN & UNWIN LTD
MUSEUM STREET

Dedicated to my partners:

FLEETWOOD PRITCHARD

SINCLAIR WOOD

E. S. DOWDALL

W. D. H. McCULLOUGH

LESLIE ROOM

GERALD BUTLER

C. C. J. SIMMONDS

BOOK
PRODUCTION
WAR ECONOMY
STANDARD

THIS BOOK IS PRODUCED IN COMPLETE
CONFORMITY WITH THE AUTHORIZED
ECONOMY STANDARDS

FIRST PUBLISHED IN 1944

PRINTED IN GREAT BRITAIN
in 11-Point Baskerville Type
BY UNWIN BROTHERS LIMITED
WOKING

Contents

Illustrations in the Text

List of Plates

Introduction

By CHARLES TENNYSON, c.m.g.

THE publication of this book seems to me most timely. It is true that the industries to which it is more particularly addressed are almost exclusively engaged in munitions production, but even those industries have to take thought for the needs of peace and many forward-looking manufacturers must already have given much consideration to peace-time products and methods. I think it must be generally recognised that the coming of peace will confront us with radically new conditions.

The world is swiftly and surely moving into a new Era, the avowed object of which will be to use all its resources, both mental and material, in a deliberate and concerted effort to increase and maintain the standard of living of the ordinary man. Everyone is concerned with the idea of planning internationally, nationally, regionally and industrially. But the idea must go much further than that if the world's new ideal is to be attained. It must get right down to the design of the ordinary products of human consumption.

Hitherto this has been largely a matter of tradition, or of rather careless adaptation to meet supposed popular taste or the exigencies of competition. What we need for the new world is a policy of scientific adaptation to need.

When I speak of design, I am not thinking of surface pattern or pure decoration. There is room for much improvement there, and from the aesthetic point of view the subject is of first rate importance, but design of that kind does not affect the utility of the product to any important degree. I have in mind the design of shape, structure and fabric, with a view to giving the consumer the greatest possible value in service, convenience and the delight of eye and touch.

To apply this principle to the vast range of articles of domestic use is a formidable task. There are relatively few which could not be substantially improved as a result of intensive and scientific study. How is this to be brought about?

Many of our industries (notably those dealing with textiles and pottery) have a long tradition of design based on study of home and overseas markets. Many large firms have well staffed and active design departments of their own. But there are many industries dealing with ordinary household equipment and requirements whose ideas on this subject have not progressed beyond the imitation or adaptation of their own and their competitors previous best sellers. There are countless firms who employ no trained designer at all and who could not afford to employ one whole time.

Right through industry there has been far too little deliberate and intelligent consideration of design. Unfortunately, this even applies to new products. Even the great new range of plastic materials had, before the war, brought very little new thought into the field.

The explanation of this backwardness is that design for industry is a very complex problem. It requires a trained creative imagination and a knowledge of machinery, materials and markets. It demands for its full realisation some control over the productive activities of the factory, to see that the designer's intentions are properly carried out, and the power to secure the co-operation of the sales force, not always an easy matter, if the design is one of any novelty. Mr. Gloag hits the nail on the head when he insists that the majority of design problems must be dealt with on a research basis and by team work, and emphasises that the designer must be brought in at an early stage of the enquiry and must be given the requisite status. He is just as much a technician and a specialist as the chemist or engineer and should be given equal influence. There are many industries where conditions make it impossible for the normal firm to afford a designer of adequate status, and even in large firms there is a danger that the design staff may become narrow in its outlook through excessive concentration on immediate problems. These dangers were recognised in America during the years between the wars and groups of Design Consultants were springing up, some with quite elaborate organisations. Part at least of the great progress which American Industry was making in design must be attributed to this development. Mr. Gloag has outlined in the Design Research Committees which he proposed, a novel approach to the problem, which has already achieved successes and may prove a valuable factor in national reconstruction. He writes with knowledge born of practical experience and with the enthusiasm of one who has for many years been in the vanguard of the struggle for better industrial design. As Chairman of the National Register of Industrial Art Designers, I know that we have in this country plenty of men and women fully competent to carry out his ideas, and I sincerely hope that their abilities may be fully utilized in the difficult years that lie ahead of us.

CHAPTER I

Why Missing ?

IN the course of a speech delivered in February 1943 on "The Future of Exports," Mr. Herbert Morrison surveyed our prospects in the markets of the world. There was a note of constructive optimism in his speech, arising perhaps from his appreciation of the scale and splendour of the opportunity that awaits British Industry. He felt—as indeed most of his fellow-countrymen must feel—that it is within our power to make the most of this opportunity, and he pointed out that in rebuilding our export trade we must exercise and improve our knowledge of market research and our skill in selling.

We must, of course, do more than study the marketing and selling problems involved; we must make certain that our goods can hold their own in foreign markets, and that they have something more to commend them than the traditional excellence of British workmanship. The reliable nature of goods turned out by British Industry is based partly on our native gift for fabricating materials—which in England was apparent generations before the Norman Conquest—and partly upon the cautious wisdom of our manufacturers. In a fairly wide experience of the workings of a variety of industries during the last twenty years, I have never met a manufacturer who would dream of putting any new line of goods or a new material on the market until he was fully satisfied that they were technically sound, and that in use they would not let down the reputation of his firm.

Hard things have been said about the intelligence of British Industry, generally by highbrow critics whose in-

tolerant contempt for private enterprise is matched only by
their ignorance of industrial operations. It is true that
between the two world wars some of the older men who
directed British Industry often had their eyes on the past:
they looked back to the days when Britain was "The Workshop of the World", when orders for goods poured through
the letter-box in a heartening stream, day after day, when
business came unsought to a firm, and few people had to
bestir themselves to woo fresh customers. There certainly was
a time when directors of many industries thought that British
goods sold themselves, or that if you had to do some selling,
it was only a matter of employing hard-drinking professional
"good fellows", who could slap a potential customer on the
back, stand him a couple and then get him to sign on the
dotted line. That generation has passed away. A new generation is now in power, mostly men who survived the 1914–18
war; and those men are not floundering wistfully in the past:
they have the vision to appreciate the scale of our future
opportunity, plus the energy and wisdom to grasp it. But
market research and selling effort for British goods are not
enough to secure and hold foreign markets; nor can we afford
to rely wholly upon our traditional reputation for good workmanship. Good workmanship is not enough. If the products
upon which it is lavished are not well designed, then it
becomes wasted workmanship.

After the second world war, Britain will have more
machines available than ever before in her industrial history.
New materials and combinations of new and traditional
materials will be available also. *How* we use these new and
immense powers of fabricating materials into commodities
will largely determine the extent of our capacity for selling
those commodities abroad. No amount of market research,
selling effort, or propaganda can win or maintain a market
for goods, if that market is invaded by producers who have
given lively and exceptional thought to the design of their
goods. We can bawl "British and Best" with the utmost

confidence in the excellence of our industrial workmanship; but nobody will "Buy British" if competitive goods are available which have greater convenience in use, are more agreeable in shape, colour and texture, and have moreover an anticipatory air of new worldliness. Such goods need not be competitive in price: attractive appearance arising from an imaginative approach to the choice and use of materials and finishes may, by comparison, make British goods look hopelessly old-fashioned. The consideration of *how* we can use our new industrial powers, our new materials, our practical skill to the best advantage, not only for our export but for our home trade, depends upon our capacity to identify and to apprehend the significance of a neglected operation in industrial production. Although they exercise wisdom and skill in planning production, many industrialists have, all unconsciously, allowed a blind spot to mar their vision. They have frequently ignored the function of design; or, if they have recognised it, they have regarded it as an ancillary instead of a basic operation.

Good design, which could become such a potent selling factor for British goods in the future, arises from the effective partnership of trained imagination and practical skill. Trained imagination gives a product character: practical skill gives it reliability.

But at what stage in industrial production does trained imagination come in?

It should come in when the industrial designer is consulted, for *he* is the man whose imagination has been trained. But when is he consulted? Is he consulted at all, and, if so, by whom? I suggest that all too often there is a missing operation in British industry, though it is one that should properly be as ordinary and as generally accepted an operation as the preparation of drawings for jigs, the lay-out of a production sequence, the testing of materials, and the rest of the normal procedure for putting some object or series of objects through production. This missing operation in industrial production is the initial consultation with the industrial designer.

The industrial designer is the missing technician in industry.

Why should such an important technician be missing? In the last century we led the world largely by virtue of our inventiveness and enterprise. At what stage in our development did the industrial designer disappear? Did he, indeed, ever emerge? In order to answer such questions and to appreciate the significance of our lost leadership in good design in many industrial activities, we must glance back at the prosperous and relatively peaceful nineteenth century.

We now live in a commercial machine age. At the beginning of that age, the chief test of success for any enterprise was its capacity to pay. It is still a salutary and effective test; for if a business cannot pay its way there is usually something wrong with its organisation or its ability for serving the needs of a market. People who were engaged in creative business long ago realised that they must accept a wider responsibility in the conduct of their commercial affairs. It was not enough for a business to earn good dividends: it must contribute something to the satisfaction and well-being of various sections of the community. This idea inspired the creators of many industrial enterprises in Britain in the past, and without such inspiration our railways and steamship lines, our shipbuilding and our iron and steel industries could not have developed, nor could the great engineers and inventors of the last century have secured the financial patronage that made their work possible. The words "satisfaction and well-being" may have a wider meaning now than men were prepared to read into them a hundred years ago; but to provide for the satisfaction and well-being of people who wanted to travel from Manchester to Liverpool or from London to Birmingham with speed, safety and comfort, was an end in itself in the early days of railways. Men built factories and mills and railways with a tremendous sense of achievement; and, proud of their various enterprises, they thought, quite rightly, of profits and of satisfying the needs of some new and growing market. By serving the market they were serving the community which was better and richer as a result of their work.

Men with a different type of creative imagination felt that all this industrial activity made the community poorer. At the beginning of the nineteenth century the Luddites were open enemies of machinery. They were machine-breakers; they were not revolutionaries, they were militant reactionaries, who wanted to restore handicrafts; but they failed to arrest the development of industry. Within fifty years of their pathetic war against machines, the Luddites had many artistic and intellectual imitators—gentlemen who were disgusted by the untidy expansion of industry. Such people were savagely critical of machines and their products. John Ruskin was their prophet. Presently they were heartened by the work of a poet and artist-craftsman, William Morris, who attempted to reinstate mediaeval handicrafts and to arouse interest in them.

"The Golden Age, up to a few decades ago, was always in the past, and many men and women, charmed by its apparent tranquillity, attempted to make their own lives independent of contemporary events; to form, as it were, an artistic and intellectual (but not an economic) oasis in the sooty desert of the commercial machine age. William Morris wrote and painted and carved and wove, inspired by the urgency of his dislike for the ugliness of nineteenth-century industrialism and his love for an imaginary paradise of mediaeval arts and crafts."*

Because thousands of creative minds were drugged by this nostalgic longing for the methods of a remote and idealised past, the development of industry was seriously

* *What About Business?* by John Gloag (Penguin Books), Chap. III, p. 25.

THE FIRST CAST IRON BRIDGE AT COALBROOKDALE

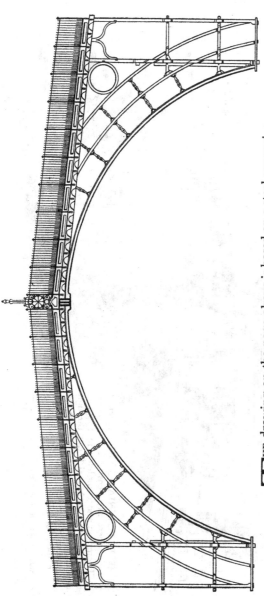

THE drawing on the page above is based on a steel engraving made in 1782, and the detailed elevation on this page is taken from an original working drawing, both in the possession of the Coalbrookdale Company who cast and erected the bridge in 1779. It was designed by Thomas Farnoll Pritchard and was constructed by John Wilkinson and Abraham Darby. It crosses the River Severn between the towns of Ironbridge and Broseley, Shropshire, and is still standing. (See page 19.)

THE BRITANNIA BRIDGE, designed by Robert Stephenson and Francis Thompson and completed in 1849. It carries trains over the Menai Strait through a double series of great box girders. In the mid-nineteenth century, it represented a new technique of design.

(This drawing is made from an engraving in the possession of the L.M.S. Railway, and is reproduced by their permission.)

injured; though the injury was at that time neither recognised nor even suspected. But during the last half of the nineteenth century and the opening decades of the twentieth, men with trained imaginations were missing from most industries. Such men had, in the early days of industrial production, been associated with industrialists. For example, in the late eighteenth and early nineteenth centuries, designers were actively concerned in extending the use of cast iron, and exploring all the possibilities of its employment on a large scale. Some of the early bridges made in that remarkable material had great elegance. At Ironbridge in Coalbrookdale, Shropshire, the first of these cast iron bridges still survives. It was designed by Thomas Farnoll Pritchard, a Shrewsbury architect, and was constructed by John Wilkinson and Abraham Darby. It was completed in 1779. Designers were then using the most "modern" material available. They were not attempting to reproduce in cast iron the characteristics of a stone or brick-built bridge: they were under no fashionable compulsion to imitate anything. Like Sir Joseph Paxton when he designed the Crystal Palace of pre-fabricated cast iron units and glass, they invented a technique for using new materials and combinations of materials. Robert Stephenson and Francis Thompson built the tubular bridge over the Menai Straits without reference to any prototype; Sir Benjamin Baker created a new machine architecture of steel when he designed the Forth Bridge.

The great engineer-designers of the nineteenth century were masters of industrial design; and on the railways they gave abundant proofs of their genius in the locomotives they designed, which had clean and beautiful lines, and reflected in their appearance their efficiency and capacity for speed. In the design of coaches they were less successful, for they were haunted by the ghost of the stage coach, and a railway coach was a series of conjoined stage-coach compartments. Even George Mortimer Pullman in America, who invented the saloon car, with its central aisle and conductors' platforms at each end, was guided in his departure from the compartment system by the form of the river steamboat. As a prototype, the river steamboat was certainly less cramping than the stage coach. Since the last decade of the eighteenth century it had grown in spaciousness and efficiency: the inland waterways of the United States had thus been developed years before railways were made. A steamboat was in operation at Philadelphia in 1790—nearly eighty years before the Pullman Palace Car Company was founded.*

Apart from these large-scale examples of the operation of design in industry, and the imaginative handling of new materials by designers, the nineteenth century missed the significance of design as an industrial operation. The designers who might have been studying mechanical processes and improving the products of industry were playing the fool with handicraft revivals, collecting and imitating antique furniture and odds and ends, chattering in studios and shuddering at the name of industry. That some men of genius were produced by the Morris school of thought is not in question: but their genius was unrelated to the problems of their time. One of the greatest of these men, Ernest Gimson, who made fine furniture by hand, realised that he had deliberately separated his work from contemporary life. He dismissed, under the wide and misleading label of *commercialism*, the possibilities and products of industry.

"He desired commercialism might leave handiwork and the arts alone and make use of its own wits and its own machinery. Let machinery be honest, he said, and make its own machine-buildings and its own machine-furniture; let

* "Near this place lay a very long steamboat, which I have seen stemming the tide at the rate of about four miles an hour."—*Samuel Kelly*. The Diary of an Eighteenth Century English sea captain, edited by Crosbie Garstin, Part II, Section 13, Entry 11th May, 1790. (Published by Jonathan Cape, 1925.)

it make its chairs and tables of stamped aluminium if it likes: why not?"*

The brains of designers like Gimson would have done better service to their country had they been even partly devoted to exploring how well industry could make furniture from stamped aluminium. Like many followers of William Morris, Gimson was an unrealist, refusing to acknowledge the fact that, for better or worse, for richer or poorer, we live in a commercial machine age. Unless we recognise the civic and commercial value of good design, we shall make this age poorer in amenities and wealth. The future development of the commercial machine age and our material capacity to make the world better or worse in the second half of the twentieth century, depend upon the outlook and actions of our industrialists and the good designers who should be in partnership with them.

How can the good designer and the industrialist work together productively?

It is the object of this book to give some practical answers to that question. They are not the only answers, nor can they possibly satisfy every type of industry, though they may suggest possibilities. The methods of collaboration I shall describe have been tried successfully in various industries, and have led the manufacturers concerned to accept industrial design as a normal business operation.

* *Ernest Gimson: His Life and Work* (Ernest Benn Ltd. and Basil Blackwell, 1924), pages 13-14.

CHAPTER II

Industrial Design as a Technical Operation

WHEN industrial design is recognised as a business necessity and practised as a normal operation in production, fresh, vigorous and progressive thinking about markets is encouraged. The recent history of industrial design in the United States is instructive. In another book I have briefly described how fully and fruitfully American industrialists have collaborated with designers.

"During the first third of the twentieth century, American industry was to attain such facility in the fabrication of commodities, that articles that were at first considered luxuries were brought within the reach of millions of consumers. During the nineteen-twenties American industry decided that efficiency in mass-production was not enough in itself. Industrialists realised that there was a missing technician. Metallurgical experts and chemists could make materials malleable and ductile; but the final form and texture of industrial products were often unconsidered, or determined by engineers whose chief concern was production, who were interested in the nature and operation of tools rather than in what those tools were making.

"At first, 'stylisers' were employed to impose a final shape on machine-made articles, to improve their proportions, and to give them a modish appearance; but this was just another manifestation of what used to be called 'applied art'. But at last the missing technician was identified, and he took his place in industry. He was the industrial designer. He was a new type of artist, prepared to study industrial production and machine processes, to investigate the properties of materials, and to relate his creative ideas to the needs of consumers. He was as much a man of business as an accountant; as much of an artist as an architect; a man who, like all real artists, identified himself with contemporary life. He took all branches of industrial production in his stride, like an artist or craftsman of the Italian Renaissance, who was prepared to be an architect, a painter, a jeweller, a sculptor, or an ironworker, as occasion demanded. The industrial designer has assumed a responsibility so wide that his ideas already influence the taste of the general public.

"In America, the man who can design a streamlined train, a motor-car body, a gas cooker, a trailer-caravan, an electric iron, or a radio set, is now regularly consulted as a technician—as a qualified engineer, a chemist, an electrical expert or an architect is consulted. The industrial designer demonstrates America's capacity for giving opportunities to the right men. Designers like Norman Bel Geddes, Walter Dorwin Teague and Raymond Loewy, are controlling the shape and character of all manner of things which affect the everyday life of the people—in their homes, in their cities and towns, and in transport."*

It should be clearly understood that the industrial designer is *not* a "styliser"; he is *not* a "putter-on" of shapes or patterns. He should be a man with an inventive and a receptive mind, with the sort of training which will enable him to apply his imaginative powers to the study of materials and mechanical processes for fabricating materials. His work begins when the production of any article is being considered, and the time to call him in is when the production engineers and the sales executives and the factory directors are making their initial plans. Leave him out of the picture, and you have left out a vital operation, and maybe you have lost the touch of distinction that would have won a new market for a product and led to big sales. I do not

* *The American Nation: A Short History of the United States*, by John Gloag (Cassell & Company), Section VI, "The Creative Contribution", pp. 372–374.

imply that production engineers, sales executives, and industrial directors are deficient either in technical knowledge, vision or inventiveness; but I suggest that their technical knowledge, vision, and inventiveness can be rendered more creatively effective if they call in an independent imaginative mind, capable of putting an independent point of view, a mind moreover that is equipped with wide knowledge of many materials and processes, and is not easily intimidated by limitations which may, in all sincerity, have formerly been accepted as customary and convenient. To illustrate this point, a specific example of the results of collaboration with a trained and capable industrial designer may be described.

Some years ago the H.M.V. organisation decided to put on the market an improved form of electric iron. When they made that decision the electric iron was rather a clumsy copy of the old-fashioned flat iron: it hardly differed from its antique prototype. It was an assembly of metal parts rather conspicuously fastened together. Now the H.M.V. organisation presented the problem of producing a convenient electric iron to an industrial designer—Mr. Christian Barman—who, after consultation with scientists and engineers, departed from the conventional methods which had hitherto been used. He designed what is now the famous H.M.V. Electric Iron, of which the visible part consists only of a completely streamlined iron without any joints, made of a material that had never been used before for such an appliance, namely, hard-glazed fireclay. The only other visible parts were the control-switch and metal sole-plate. There were no lumps, bumps, screws, or projections, or any untidy external interruptions to the surface of the H.M.V. Iron; and everybody wanted one.

That is an example of the productive use of an industrial designer. It is an example, too, of the selling power of good design. Mr. Barman, who is a distinguished architect, is one of our leading industrial designers. It is significant that the training of an architect gives the profound, far-ranging interest in, and knowledge of, a diversity of materials, which, in alliance with a creative imagination, can produce such an unconventional but highly effective and convenient article as the H.M.V. Iron. I shall return to this point later.

It is not always possible for a manufacturer to isolate a problem, or to realise that by calling in an industrial designer he may take a short cut to solving it. He may call in a designer; but he may do it too late. He may well embark upon a programme of production and be content either to allow the ultimate form of the object to be determined by the convenience of the production process, or by the ingenuity of his own production engineers, or merely allow it to take care of itself. After production is settled, he may feel that it ought to be smartened up, disguised, "stylised", or whatever he cares to call it, at a later stage, or at the final stage of production. Then he calls in a designer, too late to use his trained imagination. It is like calling in an architect to suggest a colour scheme for the outside of a well-built but badly-planned house, after ignoring the existence of architects in the original planning, and leaving everything to the builder.

The manufacturer may consult an industrial designer on one or two problems; or he may decide to employ the services of a whole-time designer. This would involve considerable cost if he employed a first-class man, for an industrial designer would be reluctant to sell his entire time to one organisation. He might agree to work for a limited period, but only in exchange for really adequate remuneration. Even so, the manufacturer is only getting the benefit of one mind. It may be a very good mind; but it may, however inventive, however flexible, tend to become stale after it has been associated for a period—say a year or two—with one particular branch of industrial production. Moreover, with increasing knowledge of the technicalities of a particular industry, such a mind may lose some of its independence; it may have some of its innovating daring unconsciously subdued. Unless an industrial designer is a most exceptional person, with great and almost explosive creative vitality, he may, if permanently attached to one organisation, tend to become too closely identified with the methods and materials of a particular production plant. To supply an industry with the continual refreshment of new ideas about design is a considerable problem; but it may be solved by setting up advisory committees for development research in design, or, as I prefer to call them, design research committees or panels.

The function, composition and operation of such committees, their periods of work, and the nature of the programmes to which they give their attention, are the main concern of the next five chapters.

The Evolution of the H.M.V. Iron

I Technically, the first electric iron represented a great advance on the charcoal and gas burning types which, in their turn, had taken the place of the old flat iron heated on a stove. But for years it did not seem to occur to any manufacturer that an electric iron might be *designed*. Its form was that of the old flat iron with an electric plug point added.

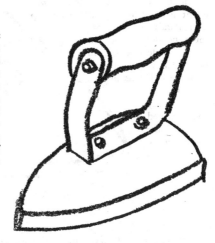

2 As the plug point prevented the iron being stood up on end, it had to be fitted with a projecting heel rest. When H.M.V. No. 1 controlled heat iron made its first appearance in 1935, the standard electric iron still looked like an old-fashioned flat iron with an elaborate contraption bolted on at the rear. It was a useful appliance, but there were some awkward drawbacks.

3 In particular, this older type was liable to become overheated and, if carelessly left switched on, might get red hot. H.M.V. made up their minds to see what could be done about it. They also decided that in order effectively to get away from the old ideas and the old standards, the clumsy and muddled forms of the past must be scrapped as well.

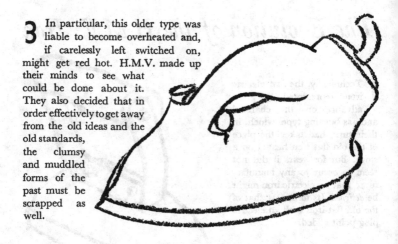

4 The company believed their scientists were capable of producing an iron in which the temperature could be regulated by the user so that over-heating would not occur. They engaged an industrial designer to help them give this revolutionary appliance an appropriate shape. The top and handle were to be made in one piece, and the usual display of visible bolts and nuts would be dispensed with.

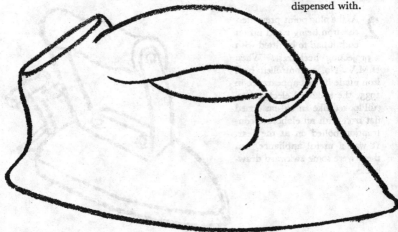

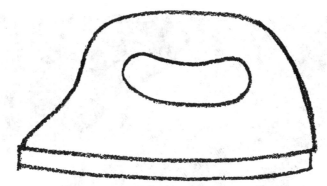

5 A small number of wooden models, built from engineering drawings showing generally the desired shape, were handled and tested by a large number of people. The results of this research were incorporated in a streamlined pattern modelled in clay. Later, casts from this model were used to continue the study of poise, plans, and easy handling. From these trials the final shape began at last to emerge.

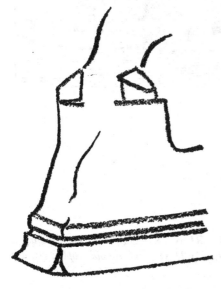

6 Simultaneously, many materials were investigated, including glass and plastics. he greatest technical problem was behaviour under heat. Once porcelain had been chosen, colour and finish were studied. Last of all came refinements such as balance and weight, a stern correctly shaped as a heel rest, and a sole plate with a sensitive nose, and an edge profiled to glide under buttons. Ambidexterous thumb rests concealed a necessary fixing.

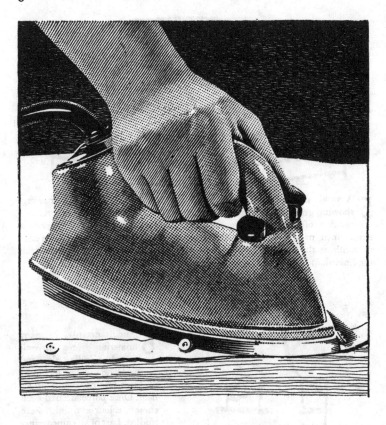

THE H.M.V. ELECTRIC IRON IN ACTION. *The drawing is
made from a photograph supplied by H.M.V. Household Appliances, and
reproduced by their courtesy.*

CHAPTER III

Design Research Committees

IN describing the character and work of design research committees, I am drawing examples from my own business experience, for my particular type of business has given me special facilities for introducing manufacturers to industrial designers and persuading them to work out problems together.

I am a director of an advertising agency. The work of an advertising agency is to introduce goods or services to various markets, and after identifying the objectives in a market, to plan a campaign of information and persuasion to reach those objectives. Markets are people: people have to be informed and persuaded. A campaign of planned persuasion may include a variety of media: press advertising, posters, booklets, leaflets, films, and sponsored broadcasting. One of the jobs of an advertising agency is the making of advertisements which appear in newspapers, magazines, and technical journals. To make advertisements, to use words and graphic illustrations in effective combination, demands different types of creative mind. There is the copy-writer who writes the words and invents ideas. There is the visualiser, who also invents ideas and who collaborates with the copy-writer, so that ultimately the right words are selected for the job, the right artist is selected to interpret the graphic ideas, and illustrations and type are allied in a given space and the result is an advertisement which appears in a newspaper.

The technique of getting artists, visualisers, and copy-writers to work together, and to work moreover with business executives who are concerned with the whole advertising

campaign and the interpretation of the sales policy behind
it, is something which advertising agencies have brought to
a high degree of efficiency. The organisation of such teams
of creative men, executives and sales experts is the day-to-
day work of an agency. Most progressive advertising agencies
are deeply concerned with the form and nature of the goods
made by their clients. Their work includes market research,
consumer research, and design research. Not only do they
make suggestions about the way goods are presented to the
consumer, in terms of packaging; but they assume another
responsibility, and occasionally give advice about the actual
design of goods produced by manufacturers. The manufacturing policy of a firm often concerns the advertising agent
as deeply as the sales policy; and this constructive and
responsible interest in manufacturing policy is not complete
without a corresponding study of industrial design. In fact,
the progressive advertising agency makes advisory work on
industrial design one of its activities.

Now a design research committee is simply a team of
creative men and executives. A technique successfully used
in an advertising agency has been applied to the problems
of industrial design and production, only instead of copy-
writers, visualisers, artists, and executives, working as
partners, there are industrial designers working in collaboration with sales executives and production engineers. Several
years before the war, in carrying out advisory work on
industrial design for a firm that was manufacturing bath-
room equipment and small articles of furniture, I helped
to organise a Design Research Committee. The work of that
design committee was the ground plan, so to speak, of a good
many similar committees which I have since been instru-
mental in forming. This committee was concerned with three
materials: timber, plastics, and glass. Plastics in those days
were rather an unknown quantity (we know much more
about them to-day), and the job of the industrial designers
was to evolve the most convenient form for a variety of

articles such as bath trays, coat hooks, and tooth-brush racks. The first drawings were made by the design members and discussed at the committee table with the plastics engineers and the sales executives. The question of the practical production of the shape suggested was thrashed out, calculations were made regarding economic runs and finishes, and the designers in consultation with the plastics technicians could, after being briefed in this practical way, modify a design and recast it to secure the best results from process and materials. At that preliminary stage, all questions arising from the initial cost of moulds, the type of plastic to be used, and the possible market, could be discussed. The drawings were then revised, and from them plaster or wood models were made, so that, in this third-dimensional stage, final adjustments and refinements could be settled. In this way, the best design for expression in terms of new and rather strange materials was secured. Before this committee was formed, it had been usual to select some familiar article made in wood or metal, and to hand it over to the plastics experts, to copy in their material.

Since that preliminary experience of a working design committee, I have, in collaboration with various manufacturers and industrial designers, set up several such committees which have addressed their attention to a number of problems. I find that manufacturers are prepared to accept the work of a design committee as a basic operation of production, as essential in its way as the research work needed for testing new materials and processes. The activities of such a committee can inject new ideas into an organisation, and a firm's permanent staff of technicians and executives is stimulated by contact with new minds, with new thoughts, and a whole range of experience drawn from other industries.

Manufacturers are not now averse to employing trained minds. Twenty years ago I used to think that the manufacturer was obstructive and difficult in the matter of industrial design. Perhaps that was true then, for the manufacturer

C

was frequently hostile to the idea of employing an "artist", and unfortunately the operation of industrial design was often presented to him in the guise of uplift. Many well-meaning advocates of betterment in design were concerned only with the "artistic" aspect of the problem; many of them assumed that business men should be only too thankful for the opportunity of learning something from the designer, and very often the designer was incapable of imparting anything except a few fierce prejudices about business. Hundreds of abortive experiments in collaboration were made, because the wrong sort of designer was selected. Sometimes designers were badly treated and exploited; very often the business men who employed them were hopelessly let down. Only when the duties and intentions of both parties were clarified by specific terms of reference could their collaboration achieve creative results.

Twenty years ago contacts between industrialists and designers were rare; even to-day they are often badly handled by both parties. The industrial designer, because he has been the missing technician for so long, is only beginning to gain the recognition and respect to which his abilities and qualifications entitle him. Unfortunately the issue is frequently confused by people who call themselves industrial designers, but who are little better than slick draughtsmen. To invite such people to cope with problems of industrial design is like asking a man who paints patterns on lampshades to do the work of an electrical engineer.

In the last chapter I referred to the significance of an architect's training in relation to industrial design. The architect is often an able industrial designer. He has the training which improves his capacity for becoming quickly familiar with a diversity of materials, and which also encourages him to be reasonably businesslike. The architect, better than any other man, can help to break down the not wholly unjustified belief of manufacturers that a designer is an unpractical visionary or a temperamental nuisance. I have found that the operation of design research committees dissipates this Victorian prejudice, for it dates from those days. The manufacturer can observe the architect or industrial designer working side by side with his own recognised technicians and executives. He is so obviously one of them. Also the architect has tangible qualifications which are confirmed by professional degrees. The qualifications of industrial designers are discussed in a later chapter, but it may be said here that the architect is not the only trained designer whose standing is indicated by professional degrees.

CHAPTER IV

Design Research Committees in Operation

ESIGN research committees perform a variety of tasks; but their work is generally related to finalising the character of articles that are to be produced by some industrial process or series of processes. A committee may give some of its time to exploring the possibilities of existing materials and processes, and may suggest new ways of using them. It may, after a preliminary period of study in close association with the technicians of an industry, tackle a programme of work. But to whatever tasks the committee devotes time and thought, its work must be directed by specific terms of reference. Committees formed because the directors of a firm have come to the conclusion that "we might as well call a few of these designer chaps in and see if they've got any ideas" are often foredoomed to futility. Designers are specialists in the production of ideas; they may be relied on for a copious flow of them; but the result may easily be an unmanageable flood of notions.

Almost anybody can have ideas: few people can make them work.

Ideas should be sought from the trained imagination of the industrial designer in relation to a specific problem. The terms of reference of a design research committee should be framed in language that makes clear the obligation of that body to produce tangible results in a specified period. A programme should be settled: a date plan for carrying it out, step by step, should be agreed.

While its composition is influenced by the nature and extent of its objectives, a design research committee must consist of two groups of people: (1) the *independent* members, who are called in by the manufacturer, and (2) the *manufacturer's* members—*specialists* and *executives* drawn from his permanent staff.

The independent members include two or more industrial designers. In addition, it may be necessary to include specialists who are not designers and who are not normally on the staff of the firm which has set up the committee.

The manufacturer would be represented by specialists in production and materials, a sales executive, and a director with powers to authorise and organise experimental work.

There should be a chairman. His qualifications must include some knowledge of or interest in industrial design, and he must also possess the ability to prevent the proceedings from becoming a wrangle between the technicians, or becoming too detailed or diffused. He must also have the strength of mind to call a halt, if too many ideas are produced, otherwise concentration on specific action would be impossible. Somebody with these qualifications may occasionally be found among the directors of a manufacturing firm. The chairmen of two of the most successful design committees of which I have been a member, have also been the chairmen of the firms who set up the committees, and who held development research work to be of such importance that they gave their personal attention to its direction. A well-organised advertising agency is usually capable of providing or recommending a suitable independent chairman, for, as I said in the last chapter, a progressive agency regards advisory work on industrial design as one of its activities.

There must be a secretary to keep the minutes of meetings and to arrange for information to be circulated. It is the secretary's responsibility to see that action is recorded. The minutes are prepared in three columns: the left-hand column is left blank for notes, the central column contains the record of proceedings and decisions, the right-hand column is headed *Action*, and in it the names of those responsible for taking action are set forth. Lucid and accurate minutes are essential, and it is a psychological point of considerable importance that the names of those committee members responsible for taking action before the next meeting should be immediately visible on each page. Copies of the minutes should be circulated to members of the committee within forty-eight hours of each meeting.

After the first meeting, the agenda arises largely from the minutes, and from the action decided on; this ensures that nothing can be shelved, abandoned or forgotten.

Design research committees are small, and their work is arranged in sessions. Generally speaking a committee has a preliminary session of six months. The preliminary session may last only three months if there are three or more industrial designers on it, that period being sufficiently long to sort out the problems and to enable specific tasks to be allotted to the different independent design members. But for a small committee six months is the most practical period for the preliminary session.

The first job of the committee is development research. The lines of development are settled at the first meeting and are agreed by the independent designers, the production technicians, the sales executive and the director of the firm.

Before the committee sits, the design members are allowed opportunities to inspect the production processes and to study the nature of the materials employed by the firm. There is a certain family likeness between the processes of industrial production: it may be said that, without attempting a facile over-simplification of the complex and infinite diversity of factory operations, that materials are generally fabricated either by the application of pressure or by being subjected to temperature changes or chemical action. These are broad classifications, and some production processes embrace them all. The study of materials is often a much wider and consequently more difficult subject than the examination of production processes. All kinds of minor operations in production arise from the need to accommodate some characteristic or idiosyncrasy of the materials used; just as the cabinet-maker, working by hand, has to allow for the softness or hardness of the wood he is using, and for the peculiarities of its grain. It is occasionally advisable to allow the design members of the committee to work individually for a preliminary period—a month or six weeks— in a purely experimental way in order to become easily familiar with the materials. During this period they should have access to the technical experts in those materials; thus they may work out experimental ideas, and become acquainted with limitations, properties and possibilities. (In later chapters, the influence of materials on design, and the results of experimental research arising from preliminary study by industrial designers, are described.) In addition to this preliminary period of investigation by the design members, it has sometimes been found desirable to allow the first session of a committee to be occupied by the discussion and examination of experimental ideas, particularly when the diverse uses of some greatly varied and comparatively new materials such as plastics are to be considered. During that exploratory first session, the design members are given a roving commission in order to solve a few problems, which they select for themselves. But when the design members have been briefed about the materials and processes, the committee can proceed to solve two or more problems of design in relation to the identified needs of actual or potential markets.

After two or three meetings, the committee acquires the collective consciousness of a team. The meetings are monthly, and their duration is anything from two to four hours; but four hours is the effective limit of a sitting. At the second and subsequent meetings, the design members and the technicians put on the table sketches, notes, brief reports, calculations and so forth, which enable ideas to be carried beyond the discussion and sketch stage to the point where they can be translated into working drawings, models and specifications. Where possible such preliminary material is sent before a meeting, by the members who have prepared it, to the secretary, who arranges for photostats, copygraphs or blue prints to be made of the sketches and for written matter to be typed and duplicated, so that copies may be sent to every member of the committee for consideration.

Apart from the monthly meetings of the full committee it may be necessary to hold informal meetings between the design members, with the chairman and secretary present, to scrutinise progress and to raise points which would only impede the proceedings of the formal committee if they were discussed there. If, for example, the design members have differences of opinion regarding the approach to some aspect of the problems they are tackling, they should ventilate them at these informal meetings. Similar meetings between the firm's technicians and the design members may be arranged to discuss various matters of detail which should not be allowed to occupy the time of the formal committee. Such contacts are arranged through the secretary of the committee. After some formal meetings the volume of work to be tackled by the design members may demand more than a monthly interval; the chairman then suggests that the interval is lengthened to six weeks or two months, arranging for informal meetings to take place during that time in order to check the pace and extent of the work in hand.

At the end of the first session of six months, the amount of work done by the committee may be reviewed and assessed. This review of progress is the responsibility of the chairman, who writes a brief memorandum on the work accomplished and the conclusions and recommendations reached through the deliberations of the committee. Both parties—the manufacturer and his technicians on the one side, and the industrial designers on the other—are now in a position to know whether they think too much or too little work has been done for the agreed cost; whether certain items on the programme are sufficiently promising to be developed in a further session, or whether enough work has been advanced to the experimental stage to justify suspending the committee's activities for three or four months, while the practical job of translating its work into production is carried out in the factory.

The chairman, after consultation with the directors of the firm, may recommend a temporary cessation of meetings, while one or more of the design members are given some particular task to work out over an agreed period, or while some investigation is conducted, the need for which has become apparent during the first session. For example, some time ago a certain firm concerned with the large-scale fabrication of various types of plastics, decided to form a design research committee. At an early stage it was realised that much fresh experience was being gained in the manufacture and fabrication of new plastics in the United States, for America was not then at war. One of the members, who was not a designer but had special qualifications as a consultant on industrial design, visited the United States, and over a period of seven months made a thorough investigation of the American plastics industry. This investigator returned with an invaluable and highly detailed report, and a comprehensive collection of samples. When that particular committee resumed its sittings, it was invigorated by a new knowledge of practical possibilities.

When a committee meets for a second session it is obviously better qualified to tackle a programme of work than when it first assembled. Its composition may be changed; a new design member may be added, or one of the original members replaced. As it should be stated, beyond the possibility of misunderstanding, when committees are formed, that their work is to be carried out in sessions, and the engagement period for designers is thus clearly defined and confirmed in writing by both parties, there need be no hard feelings about changes in the composition of the committee after a session is completed.

The life of a committee may last for three sessions, if there is a large programme of work. After two or three sessions, individual design members may undertake special tasks, completing them when the committee has been wound up, and working as independent designers for the firm. Design research committees may be planned for one introductory session of three or four months, with the deliberate intention of using that session as an opportunity for briefing the design members, before giving each of them some particular job to carry out, or some line of research to explore.

From the operation of a design research committee a firm acquires the technique of employing designers productively. The firm widens its contact with and understanding of industrial designers, and the establishment of a design research committee may often lead to the formation of a permanent advisory committee on design, working in the closest collaboration with the marketing department. Standards of performance have been established; the firm knows what to expect, and its directors know exactly what it costs to buy the advice and ideas of capable and qualified designers. Illusions and prejudices about the economic place of industrial design are dispelled when its practical significance as a normal business operation is thus demonstrated and accepted.

CHAPTER V

Costs, Royalties and Patent Rights

GOOD design has to be paid for; the advice and work of able men with trained imaginations cannot be bought cheaply. The setting up of design research committees facilitates a frank and fair discussion of fees, and after the preliminary session readjustments may be considered, if desired, by the independent design members or the firm employing their services. It may be difficult for the directors of a firm to decide upon what scale industrial designers should be remunerated and by what method; but the problem of assessing the potential value of their creative work is not insuperable. The problem must be approached with faith in the capacity of the designers; without that element of goodwill, no design committee can possibly operate productively.

Mutual respect for and understanding of each other's qualifications and functions are essential if the manufacturer and the industrial designer are to enjoy an effective partnership. I emphasize this point particularly, because appreciation of its significance should inform both parties when they make their initial bargain. It may seem an obvious point; but its importance is not always apprehended by designers, though manufacturers, when they have become convinced of the need for development research in design, are prepared to back their conviction with faith in the undisclosed skill of the designers they have decided to employ. It is not always easy for a manufacturer to generate such faith, for he has a perfectly natural tendency to think that his particular form of industrial production has unique features which make his problem of design quite different from all others, that an outsider, an independent technician, cannot possibly master its intricacies, and that principles which may be valid for other industries are inapplicable to his own. The technique of the design research committee ultimately changes this point of view; but before a committee is in operation, when the cost of running it is still under discussion, the manufacturer has to make up his mind to risk time and money on an experiment, the results of which cannot be measured in advance, although the operational costs may be estimated.

The importance of making a proper bargain, satisfactory to both parties, cannot be over-emphasized. A few months before this book was written, I gave an outline of the nature and work of design research committees in a paper entitled *Design for To-morrow*, read under the auspices of the National Register of Industrial Art Designers.* In the discussion that followed, questions were asked which suggested that some designers believed that they were often exploited by manufacturers. This is largely the fault of designers: it is their responsibility to see that they receive a proper fee, unless they delegate responsibility for negotiating that fee to a third party. But here again, the technique of the design research committee permits the re-examination of fees after a preliminary period of work has been accomplished, so that any dissatisfaction may be ventilated and readjustments proposed and discussed. Designers are spending time and drawing on accumulated skill and knowledge when they set out to solve a problem. It is for that skill and knowledge that fees are paid. The value of creative work in design cannot be assessed by the amount of time expended by the designer. This point was made dramatically and memorably by a great artist sixty-five years ago. In the case of Whistler *versus* Ruskin, Whistler admitted under cross-examination that his works sometimes took him very little time to execute. The picture over which he had brought a libel action against

* Read at the Royal Society of Arts, March 23rd, 1943.

Ruskin only took two days to paint. "The Attorney-General demanded if it was for two days' work that Whistler asked two hundred guineas, and the artist retorted: 'No; I ask it for the knowledge of a lifetime'."*

An industrial designer must realise that much of his work is experimental. Although many of his designs may be tried out, not all of them will be workable or acceptable in the preliminary session of a design committee. Against the good ideas he initiates, which ultimately pass into production and secure satisfactory sales, must be offset the experimental research costs incurred on the dozen or so ideas which may have been unsuccessful, or which at least were unproductive of early economic results.

That the eventual results of the industrial designer's work may be economically justified has been proved by many firms; and to bring the whole question of fees into its proper perspective, I am quoting some statements made by Mr. Raymond Loewy, the American industrial designer, in a paper on *Selling Through Design*, which I read on his behalf, before the Royal Society of Arts in 1941.† Mr. Loewy's paper was written several months before America entered the war, and after a compact and excellent sketch of the growth of industrial design in the United States, he gave some specific facts and figures concerning the range of his own work. He anticipated that during 1941 some $850,000,000 worth of manufactured goods would appear according to his design specifications.

"For General Motors", said Mr. Loewy, "I design the complete line of Frigidaire products—ranges and refrigerators, and whenever I say 'I design' I wish it to be understood 'in collaboration with the very capable engineers of my client companies'. In 1940 an increase of 100,000 units sold—

* *Whistler*, by James Laver (Penguin Books edition, 1942), Chap. VII, p. 103.

† December 3rd, 1941. The paper is printed in full in the *Journal of the Royal Society of Arts*, No. 2604, Vol. XC, January 9th, 1942.

i.e. 25 per cent more than the previous year—is attributable in part to design, and this for a product which has been a leader in the field for years. Studebaker's increase in sales of about 128 per cent from 1938 to 1940 coincides with my retention by that corporation."

Mr. Loewy also dealt with the operation of industrial design in relation to such large-scale problems as railway rolling stock. He used the experience of the Pennsylvania Railroad as his principal example. Following the redesigning of coaches, sleeping, lounge and dining cars, he records that passenger traffic on a particular run increased by 37 per cent over the previous average within a year, following the introduction of the new train. "Public prestige increased for the company; this alone is worth millions of dollars to the management. Little by little the railroad company started the complete modernisation of all its equipment. Ancient, wooden stations were anachronisms in comparison with the modern trains. The transfer from the design of rolling equipment to stationary units of the system was natural."

For the redesigning of small stations on the Pennsylvania system, Mr. Loewy developed "a type of pre-fabricated structure which can be constructed in series and moved to the desired location by rail, completely assembled. The flexibility and economy incorporated in this method of construction is a contribution of the designer. His experience in connection with many types of problems gives him an advantage which he passes on to each new client. Again, for Pennsylvania Railroad, an extensive modernisation of all their city ticket offices has been launched. In all these assignments product design is only a subsidiary of the over-all design. All fixtures, furniture, silverware, fabrics, display stands, floor coverings are special product designs for the whole. These are no longer classed individually, but are included in a project which is noted as 'a design'—singular! In the design of ocean liners the same thing applies".

In England, Mr. Brian O'Rorke has shown by his work

on the Orient liners, *Orion* and *Orcades*, how the whole character of a ship, apart from its equipment and interior decoration, derives benefit from the skill of an able industrial designer, working in collaboration with naval architects. The extreme popularity of those liners and the immense additional prestige they brought to the far-sighted company who built and put them into service, economically justified such a large-scale experiment in industrial design. (See Plate I.)

It is difficult to form in advance even an approximate idea of the extent to which the setting up of a design research committee may be economically justified. Its costs may be related to the final results if some royalty basis is agreed as part of the designers' remuneration; but there are practical objections to the payment of royalties on products that have been designed as the result of deliberations by a committee. Royalties may be paid on the work of individual designers, when they have initiated and carried through a particular design to the point of actual production; but negotiation for such a form of payment should be undertaken only after a manufacturer has had the opportunity of working with a designer. The design research committee furnishes this opportunity, and generally the best time for considering a royalty basis is when the sessions of a committee are concluded, and its work is carried on by individual designers, each concentrating on a special problem. The fees agreed for a designer's work on the committee will cover every activity embraced by the agreed terms of reference. If the manufacturer commissions a designer to do any specific piece of work outside the committee's terms of reference, that must be the subject of separate negotiation.

If a designer puts forward any patentable design or device arising from his work on the committee, this may be registered by the manufacturer without further obligation to the designer. There must be no doubt about the manufacturer's right to patent such designs or devices: they are occasionally produced as a result of the interchange of ideas between the design members and the technical members of a committee. To identify the originator of the idea from which some patentable design is evolved is difficult when a team of people with lively and inventive minds is at work; moreover as the technicians employed by the firm are generally under some agreement whereby they relinquish rights to any patents that result from their work, an invidious distinction in favour of the industrial designers might seriously disrupt the work of the committee. The design members must accept this provision about patent rights for their work. When they are working individually for a firm, after a design research committee has been wound up, some other arrangement may perhaps be considered, which would allow them to retain or share patent rights arising from any design they produced. It should be realised that the work of an industrial designer is so closely associated with that of the technicians employed by a firm, that it would not always be possible for him to claim responsibility for every detail of a design: modifications, improvements, the use of this or that material, may all be prompted by consultation with production engineers. We may repeat Raymond Loewy's frank admission: "Whenever I say 'I design' I wish it to be understood 'in collaboration with the very capable engineers of my client companies'."

The fees to industrial designers for their work on a design research committee must adequately cover the production of original ideas, the manufacturer's right to patent anything arising therefrom, and the specific tasks assigned to the design members of the committee in the agreed programme. If, as often happens, the setting up of a design research committee is recommended by the firm's advertising agency, the preliminary negotiations with appropriate industrial designers may be conducted on the firm's behalf by the agency, and a complete schedule of costs worked out. This schedule should cover the following costs:—

Fees to industrial designers.

Fee for providing industrial design consultant to act as chairman, who will be responsible for organising the meetings of the committee and seeing that the programme of work is carried out.

A reserve to cover the cost of secretarial work, and for incidental expenses, such as the duplication of drawings, sketches, and memoranda.

A reserve to cover drawing office costs. It must be decided who will accept responsibility for producing working drawings; whether the industrial designers are to furnish facilities for this, or whether the manufacturer will allot a draughtsman or draughtsmen to deal with this part of the committee's work.

A reserve for experimental models, materials and labour. (This expenditure can only be roughly estimated at the outset by the chairman, who is planning the work of the committee, after consultation with the manufacturer.)

A general reserve should be set aside to cover unforeseen expenditure. For example, it may be necessary for tests to be carried out to explore some hitherto untried use of an old or new material. This need, impossible to anticipate, may arise in the course of the committee's work.

The schedule of costs should cover a session of the committee, and this should be for the agreed initial period of, say, six months. Fees, and running costs, covered by the various reserves, are charged monthly, the industrial designers rendering their accounts direct to the manufacturer, and the advertising agency charging its fees for the work of the chairman and the secretary.

Industrial design consultancy work is undertaken by advertising agencies on a fee basis; familiarity with the products and sales problems of their clients gives them a practical aptitude for such work. The organising and running of a design research committee for a client, and all the initial advisory work it entails, does demand a wide knowledge of contemporary industrial design and familiarity with the work of capable designers. Nearly all my own practical experience of industrial design consultancy work during a

D

period of some fifteen years, has been acquired within the framework of an advertising agency; and this has led me to believe that the progressive advertising agency is well qualified to advise its clients upon industrial design and the selection and employment of designers. In conducting its own business, an agency acquires considerable experience of organising the production of creative work and choosing appropriate specialists in different departments of industrial design and commercial art.

CHAPTER VI

Selecting Designers

THROUGHOUT this book I am dealing only with indus-trial design and not with commercial art. The latter activity is concerned largely with the distribution of goods; its principal manifestations are in press advertising, posters and the display and packaging of goods. Despite the variety of its forms, and the considerable scope and oppor-tunity it allows to artists who can express ideas graphically or who possess an aptitude for inventing patterns and surface decoration, commercial art does not present such a range of complex problems as industrial design.

I am not dismissing commercial art as unimportant, nor suggesting that artists whose gifts find satisfactory creative and economic expression in this field are incapable of employing those gifts as industrial designers. Ten years ago in my book *Industrial Art Explained* I said: "At present a far greater number of competent designers find employment in connection with distribution than with production. Adver-tising has provided abundant and encouraging patronage for design. Commercial art is healthy and vigorous and has attracted some first-class talent. When industrial art attains the same standards of health, and encourages its practi-tioners to develop the questing, experimental outlook upon their work that distinguishes the men who do the best work in advertisement design, we shall live in a new age, com-parable in achievements, in orderliness and universal come-liness to that golden age of design, the eighteenth century—but far more exciting."*

Up to the outbreak of the second world war it was still

* *Industrial Art Explained*, Chap. VII, p. 174.

true that more opportunities for creative work by designers were provided by commercial art and by what may be called industrial decorative art: the technique for employing designers was known and practised. In making what may seem to be an arbitrary distinction between industrial design, commercial art and industrial decorative art, I am merely using a convenient method adopted by the National Register of Industrial Art Designers, who use two broad, main headings for classifying the work of their designers: (1) Designers of Shape; (2) Designers of Decoration.

The technique of the design research committee is, generally speaking, suitable only for designers of shape: surface decoration, colour and ornamental patterns are not ignored by such committees; those matters are not their primary concern: they represent one stage only of their work. A problem that is one of decoration alone usually demands a personal solution: it is essentially a matter for the direct application of an artist's inventive fancy, guided by study of the process which will reproduce his decorative patterns. For example, a firm of potters like Josiah Wedgwood and Sons, Limited, has a profound experience of finding and employing artists, dating from the eighteenth century. That great industrial genius, the original Josiah Wedgwood, knew how to discover and train artists; and he commissioned some of the best talent of his day. In 1775 he hired John Flaxman, the famous modeller: in 1782, he engaged Henry Webber on the recommendation of Sir Joshua Reynolds and Sir William Chambers. Wedgwood was in personal touch with great painters and architects: he was a member of the Royal Society of Arts, which included in its membership architects like Robert and James Adam and Sir William Chambers; painters like Sir Joshua Reynolds, the First President of the Royal Academy, and Benjamin West; artist-craftsmen such as Thomas Chippendale, the furniture-maker; John Baskerville and William Caslon, the typefounders; Thomas Grignion, the clockmaker; and engineers, like Robert Mylne, F.R.S., who had designed Blackfriars Bridge and was engineer to the New River Company. Josiah Wedgwood played an active and influential part in contemporary taste: he was not separated from the artistic thought of his time; he helped to form it; he collaborated with artists, and was an excellent and enlightened employer. "The young artists, modellers and sculptors discovered and encouraged by premiums from the Royal Society of Arts were employed or commissioned by Wedgwood to do artistic work for him at Etruria."* To his partner, Thomas Bentley, he once wrote: "You must be content to train up such Painters as offer to you, & not turn them adrift because they cannot immediately form their hands to our new style." Mr. Ralph M. Hower in his compact and informative essay, *The Wedgwoods: Ten Generations of Potters*, points out that "Without doubt Wedgwood had good taste and a real appreciation of artistic work, but it is equally certain that he seldom forgot that he was in the business to make money. For example, he restricted the production of fine pieces of ornamental ware in order to avoid spoiling the market. With useful ware, however, he saw the advantage of producing standard designs and tried to avoid special orders. Referring to one such order he wrote: 'I could sooner make £100 worth of any ware in the common course that is going, than this one sett. It is this sort of *time loseing* with *Uniques* which keep ingenious Artists who are connected with Great Men of taste, poor & wod make us so too if we did too much in that way'."†

Wedgwood appreciated the promise of the market that awaited mass-produced goods. The firm he founded enjoys

* *Pottery in England's Industrial History*. A lecture delivered before the Royal Society of Arts, by John Thomas, M.A., Ph.D., February 12th, 1936.

† *The Wedgwoods: Ten Generations of Potters*, by Ralph M. Hower, Part I, 1612–1795. *Journal of Economic and Business History*, Vol. IV, No. 2, February, 1932, pp. 296–297.

a tradition of leadership and skill in decorative industrial art: they have recently employed artists of the standing of the late Eric Ravilious and industrial designers such as Keith Murray. Other firms have shown a comparable understanding of the commercial value of using the work of contemporary artists. For instance, a few years before the second world war, E. Brain and Company, Limited, of Stoke-on-Tre it, commissioned several well-known artists to produce decorative designs for china tea-services and earthenware dinner-services, namely: Paul Nash, Sir Frank Brangwyn, Dame Laura Knight, Ernest Proctor, Mrs. Dod Proctor, Duncan Grant, Vanessa Bell, John Armstrong, Ben Nicholson, Barbara Hepworth, Allan Walton, Albert Rutherston, Graham Sutherland, John Everett, Milner Gray, Moira Forsyth and Gordon Forsyth. There are many such examples of collaboration between artists and manufacturers: Paul Nash has designed patterns of glass for Chance Brothers and Company, Limited, and printed linen for the Calico Printers' Association. But nearly all these examples of effective collaboration are concerned primarily with decorative industrial art, with the use of colours and patterns on surfaces, with the study of textures.

I am not belittling the skill and genius of artists whose gifts are thus aptly employed by industry; but I suggest that additional skill, and sometimes a different type of imagination, are needed to explore such problems as designing domestic utensils, kettles and saucepans; cooking apparatus, gas, electric or solid fuel-burning stoves; heating or lighting units; motor-car bodies; lawn-mowers or trailer-caravans. The industrial designers who tackle these problems in collaboration with production engineers must possess not only imagination, but the capacity for studying the processes and materials involved, and apprehending their possibilities and limitations.

How can the manufacturer know that he is choosing the right designers?

The formation of a design research committee is a method that ensures an early opportunity for the manufacturer to become familiar with the aptitude of the industrial design members. In the first place he takes their qualifications on trust; and it is essential for the manufacturer to appreciate the significance of those qualifications. As the importance of industrial design gains wider recognition, and its potency in industrial production becomes apparent, many alleged "specialists" will inevitably start various "rackets" for selling the work of designers to manufacturers, keeping the designers themselves in the background and presenting their work anonymously. These "back-room boys" may be hack draughtsmen, or, even more dangerously incompetent, they may be slick draughtsmen, disguising their inability to design by their capacity to produce modish drawings, coloured perspectives of "streamlined" shapes. The ability to draw does not confer the ability to think; nor does skill with a pencil automatically imply a capacity for studying and solving problems of industrial design. The technique of the design research committee protects the manufacturer. It is vital for him to have direct and continuous contact with the designers he employs. He should never accept anonymous work, purveyed by any organisation purporting to advise upon industrial design. When he does so, he may well become a victim of super-salesmanship, practised by unscrupulous impressarios of unknown and unqualified "back-room boys" whose dubious productions bear about the same relation to good design as the coloured water sold by street-corner quacks bears to the treatment prescribed by intelligent and fully qualified physicians.

In Chapter III, I touched on the value of the architect's training as a preparation for practising industrial design. The exacting demands of that training, the examinations an architect has to pass before he can qualify, and the problems he solves in the course of practising his profession, discipline his imagination, and encourage the habit of lucid thinking.

The architect is one of the few members of the community who is trained to think logically, step by step, for he has to plan buildings, to foresee all the sequences of building operations, so that the various trades do not clash, time-tables are kept, and estimates are not exceeded.

The industrial designer who is an Associate or a Fellow of the Royal Institute of British Architects may be accepted without hesitation as a man with tangible qualifications, earned by hard and continuous study. The possession of those qualifications does not, of course, guarantee the possession of imagination, though architects deficient in that quality often lack the enterprise which would lead them to explore the subject of industrial design.

I do not suggest that only architects may be employed. Industrial designers may have other qualifications. In 1936, the Board of Trade set up the National Register of Industrial Art Designers, which began work early in 1937. At the present time,* there are some 731 registered designers whose work reaches the high standard required by the sectional adjudicating committees of the National Register. These adjudicating committees scrutinise all applications, and only applicants of whose capacity they are satisfied, by an exhaustive examination of their work, are allowed to become registered. As I have said earlier, the National Register has a series of classifications, divided broadly under the two main headings: (1) Designers of shape; (2) Designers of decoration.

Under the first heading, *shape*, there are such classifications as the following, though this is not a complete list: architectural metal-work; kitchen, sanitary, heating, refrigerating and lighting equipment; ceramics of all kinds; domestic glass; architectural glass; metal stampings; light metal castings; furniture; cutlery; electro-plate; silversmithing, etc.; electrical apparatus, hot plates, vacuum cleaners, and so on; also leather goods, medals, coins, jewellery, dress

* Autumn, 1943.

accessories, radio apparatus, toys, and various other classifications.

Under the second heading, *decoration*, many categories are included, such as wallpapers, textiles, carpets, rugs and so forth, decorated pottery and tiles, lamp shades, marquetry and inlay, etched and engraved glass. There are other sections, including commercial art, but more particularly exhibition design, display and packaging; also interior decoration, and book production, bookbinding, book jackets and book illustration.

It is significant that architects are only admitted to registration if they design for mass production, but the architects who are registered include some of the best industrial designers in the country. A designer who has been accepted is allowed to use the initials N.R.D. after his name. Registration is open to any qualified designer who is a British subject, or who, not being a British subject, ordinarily works and resides in the United Kingdom, and whose work in the opinion of the governing body, and its advisory committees, reaches the required standard of design. The governing body of the National Register consists of three groups with equal representation, nominated by:

1. The Council for Art and Industry.
2. Art Societies to represent design.
3. Trade associations to represent industry.

The Associations and Societies represented are as follows:

Federation of British Industries.
Association of British Chambers of Commerce.
Royal Academy of Arts.
Royal Society of Arts.
Design and Industries Association.
National Society of Art Masters.
Conference of Central Art Institutes of Scotland.

The National Register of Industrial Art Designers is one of the manufacturer's safeguards against bogus work. The letters N.R.D. may be trusted to stand for competence, and, very often, for considerable experience. There are other significant qualifications and distinctions.

The Royal College of Art issues diplomas, and the Diploma Course is planned for students who spend three full years at the College. The Diploma of Associateship entitles a student to use after his name the letters A.R.C.A. There is a school of Design at the College. A very high standard is required from students who enter the College, and it may be said that the cream of the art schools of the country sit for the competitive Entrance Examination.

In 1937 the Council of the Royal Society of Arts made an Ordinance with a view to encouraging a high standard of industrial design. In order to enhance the status of industrial designers, it singled out for special distinction, designers who had attained considerable eminence in creative design for various branches of industry. The Ordinance allowed the Council, in their discretion and on behalf of the Society, to confer upon British subjects who had attained to high eminence and efficiency in creative design for industry, the title of "Designer for Industry of the Royal Society of Arts". Every person upon whom this distinction is conferred is given a Diploma, issued under the authority of the Council, and may use the letters R.D.I., after his name. The number of persons who bear the distinction of Designer for Industry on the Society register at any time must not exceed 40.

The qualifications of some of our leading industrial designers are instructive. Many of them are either architects or have had architectural training. Here are a few examples which illustrate the diversity and range of qualifications:

Christian Barman, F.R.I.B.A., N.R.D.
Wells Coates, B.A., B.Sc., Ph.D., F.R.I.B.A., N.R.D.
Frederick Gibberd, F.R.I.B.A., N.R.D.
John Grey, F.R.I.B.A., N.R.D.
Keith Murray, F.R.I.B.A., R.D.I., N.R.D.

A. B. Read, A.R.C.A., R.D.I., N.R.D.
Brian O'Rorke, M.A., F.R.I.B.A., R.D.I., N.R.D.
Gordon Russell, R.D.I., N.R.D.
R. D. Russell, N.R.D.
Grey Wornum, F.R.I.B.A., N.R.D.

In this country we have some of the finest industrial designers in the world, a fact that is not always appreciated. We have a regrettable habit of thinking that somebody with a foreign name must have some highly specialised knowledge about design, something modish, something fashionable, something which no Englishman can ever have. That state of mind is due largely to the admirable propaganda organised by Colbert, Louis XIV's great Minister, who, over two and a half centuries ago, set out to impress upon the world that the arts and manufactures of France were supreme in modish elegance. The propaganda was successful: the belief remains.

We have in the interval between the wars, often displayed a weak distrust of our national capacity, not only in matters of design. I repeat, that we have no need to seek abroad, in Europe or America, for industrial designers. In many fields, the technique of the design research committee has proved to industrialists the practical capacity of their own countrymen in design. The more people, the more firms, who are working out these and other methods of collaboration with industrial designers, the better for the future of British industry. We must tap the great reservoir of talent which exists in our own country. I have named only a few of our industrial designers: in this book I have illustrated some examples of their work. Men like these can give such distinction and such variety to British goods of every description, that although Britain in the years to come could not hope again to be called the "Workshop of the World", she might well hope to maintain undiminished her reputation for magnificent workmanship and to add to it a new creative leadership in industrial design.

CHAPTER VII

Examples of Work by Design Committees

HAVING discussed the operation, composition and costs of setting up and running design research committees, some examples should be furnished of the work they undertake. It is not possible to describe in detail the work of any actual committees, because their proceedings are confidential; but some broad indication of activities may be given. Often during the war period the work of a design research committee has been concerned with specific war and post-war problems of design. For instance, one of the most pressing and urgent of post-war needs will be the speedy provision of considerable numbers of houses. Some time ago a design committee was appointed by a firm of manufacturers of walling material. This particular firm had considerable experience in the production of standardised units for various kinds of partitioning for internal use. The design research committee was instructed to examine the large-scale application of such pre-fabricated units to wartime shelter and possibly to post-war housing. The design members consisted of two architects and a woman specialist in domestic science. The firm provided a technical expert in the production of the material; a structural engineer; a sales director, and the managing director. I was chairman, and one of my assistants acted as secretary.

The first session of this particular Committee lasted for six months. During that period there were six formal meetings. Before the session ended it was decided that the designing of independent pre-fabricated units did not go far enough, and that an experiment must be made with a complete house. The life of the committee was renewed for a further session of six months and the production of detailed plans for a house was aimed at, and a system was evolved for pre-fabricating the framing for such a house. It was at this stage that, in my capacity of chairman, I was able to associate this particular committee with the work of another design research committee which I had been organising for another firm. This second committee was particularly interested in roofs and in certain internal forms of metal framing. The work of the first committee was carried to the stage when its findings could be translated into action by the firm that had convened it. Both firms now decided to form jointly a small research group, in order to solve some of the problems of pre-fabricated house construction. Other organisations became associated with the work of this group, and their research assets in design were pooled. The erection of experimental houses, duly sanctioned by the proper authorities, was the stage that followed. These are two examples of the work of design research committees which have been mutually stimulative, and have resulted in productive collaboration not only between manufacturers and designers, but between manufacturers of complementary building materials.

The second committee, which was studying roofs and forms of internal framing for houses, had for its design members three architects, two of whom were industrial designers, and a civil engineer; the convening firm was represented by two production engineers, a marketing executive, and a director in charge of development.

Now here is an indication—necessarily sketchy—of the proceedings of a design committee. Let us suppose the programme has been agreed and that the committee is concerned, shall we say, with examining the possibilities of a cheap type of kitchen shelf-and-storage unit, incorporating a refrigerator, which could be standardised for use in small houses; though that, of course, is only one item on its programme. I should make it clear that this is a hypothetical case, but it bears a resemblance to many of the problems which have been solved by this technique of design research. Let us suppose, then, that the committee is set up by a firm with facilities for woodworking, for pressing and casting metal and moulding plastics, that they have also facilities for assembly, and can put into production any number of objects, from radio sets to washing machines, wheelbarrows, or institutional furniture. The firm's team would consist of two of their own technicians, acquainted with production methods for wood, steel, aluminium, cast iron or plastic objects; a refrigeration expert, who has dealt with that particular side of the firm's work, for they have experience of the refrigeration business, though only in the high-priced end of the market; and the director in charge of development who would represent the sales side of the business. The design members would include two industrial designers, who are also architects, and a woman expert in domestic science. There would be an independent chairman and a secretary. To clarify the imaginary proceedings of this committee, we will label the different members as follows:—

Chairman

Design Members	Manufacturers' Members
D1 Industrial designer	M1 Development Director
D2 Industrial designer	M2 Production Engineer
D3 Domestic Science expert	M3 Materials specialist
	M4 Refrigeration Expert

Secretary

We will assume that we are attending the second formal meeting of the committee. The minutes of the previous meeting have been circulated, and the Chairman signs them. The agenda of the meeting arises largely from points in these minutes, for example, the first item may be:—

Preparation of preliminary sketches and notes to show how the projected refrigerator forms a basic standard unit, to which other units may be added, so that a complete "Kitchen-Set" may be evolved.

In the *Action* column of the Minutes the names of D1 and D2 are entered for sketches, and D3 for notes.

It was arranged that D1 and D2 should complete their sketches ten days before the second committee meeting at which they were to be discussed, and they were sent to the secretary, who arranged for photostat copies to be circulated to all the members. This has allowed members a few days in which to form ideas and prepare points for discussion at the meeting. Under this item on the agenda, the Chairman, therefore, invites D1 and D2 in turn to amplify verbally any points which they may wish to make about their sketches. He also invites D3 to give her views. The sketches have been prepared after an informal meeting between D1, D2 and D3. Views are then sought from M1 who may express an opinion from the sales angle. M4, the refrigeration expert, will then say how well, or otherwise, D1 and D2 have interpreted the technical brief which he provided for their guidance. M2 and M3 deal with the practical considerations of producing units of the kind indicated in the rough sketches.

This is an attenuated outline of the type of discussion that should take place; but it is the Chairman's job to see that all the members keep to the point and that no idea is pursued too long, unless there is some prospect of it being effectually overtaken, and productively employed to advance the work in hand. It may seem a relatively simple matter, and one hardly justifying much preliminary discussion and experiment, to design kitchen storage units which incorporate a standard refrigerator; but many factors have to be considered, such as the relation of such units of internal equipment to standard kitchen plans in different types of houses, the possible variation of the units, so that their convenient disposal in a kitchen may be facilitated, and the maximum convenience attained. All these matters are touched on and ultimately examined by the design members, if they cannot actually be finalised at a committee meeting.

Agreement having been reached, regarding the principles indicated by the rough sketches, a time-table is then worked out, which allows for the preparation of working drawings by the technical staff of the firm, so that a mock-up of a series of units may be made and inspected. At this second meeting of the session, the whole of the future of this particular design, which represents one item on the agenda, should, as far as possible, be foreseen. A progress plan should be agreed so that the approximate date when it may be possible to put the design into production may be reasonably estimated after all the intermediate stages have been measured and due allowances made for unforeseen delays. A time reserve on the work of a committee is just as necessary as a cash reserve to cover unpredictable expenditure. Such timed, working plans would be the aim; it should be the task of the chairman to impress upon all parties concerned the importance of making and keeping to such plans. Although many ideas are exchanged at formal committee meetings, the chief reason for them is to make arrangements for getting the work done, to place responsibility on the people who are to do it, to make sure that no friction has arisen between any members of the committee, and to establish beyond dispute that everybody clearly understands what they are supposed to be doing and why. Thus the minutes of each meeting are an inescapable record; and it is important for a secretary with executive capacity to be employed. Her work does not end with the recording of minutes: she has to watch, control, and tactfully expedite

TWO EXAMPLES OF THE WORK OF A DESIGN RESEARCH COMMITTEE

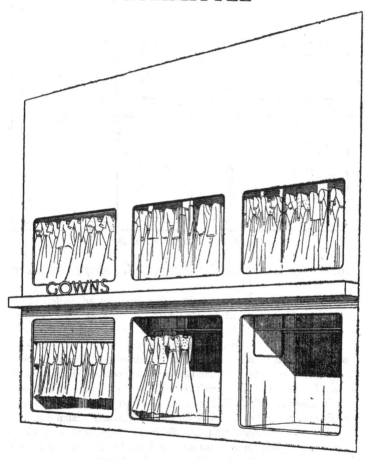

ADAPTABLE SPACE SAVING UNIT
(Description and detailed drawings on pages 66 and 67.)

E

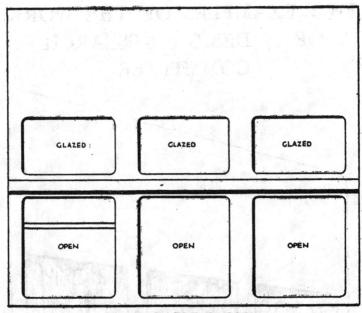

ELEVATION TO SHOWROOM

This space saving unit is designed for shops and stores

This unit consists of a series of six rails carried on an endless chain, operated by a small electric motor which can be worked by the press button method, so that any section or layer can be sent direct to the serving floor. It enables any specific section or layer to be directed to the sales floor for goods to be taken from it; at the same time, the layer immediately above is visible, so that two sets of goods are simultaneously displayed. (See perspective sketch on previous page and sectional plan and elevation opposite.) The unit was evolved by a design research committee appointed by Harris & Sheldon Ltd., whose design members were Christian Barman, F.R.I.B.A., Brian O'Rorke, F.R.I.B.A., Grace Lovat Fraser, Jane Drew, F.R.I.B.A., and R. D. Russell, N.R.D., the technical members being E. C. Stewart and A. W. Shearer.

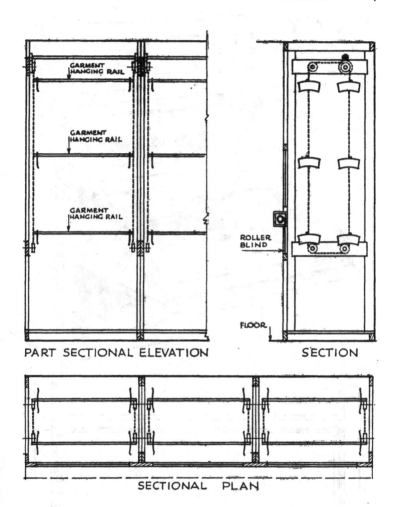

PART SECTIONAL ELEVATION   SECTION

SECTIONAL PLAN

ADAPTABLE SPACE SAVING UNIT

(See pages 65 and 66.)

T R A Y S

with disappearing
flap for various types
of storage units.
(Sectional plan be-
low; sections show-
ing mechanism on
opposite page.)

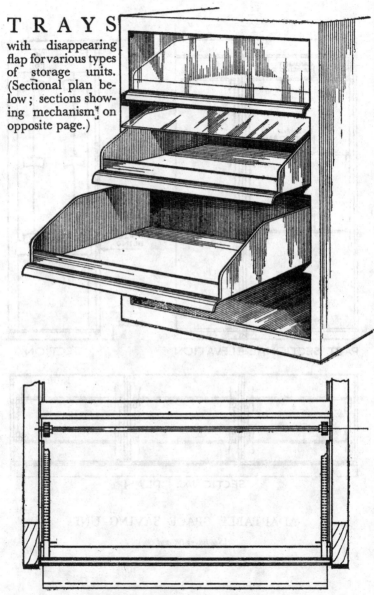

PART SECTIONAL PLAN

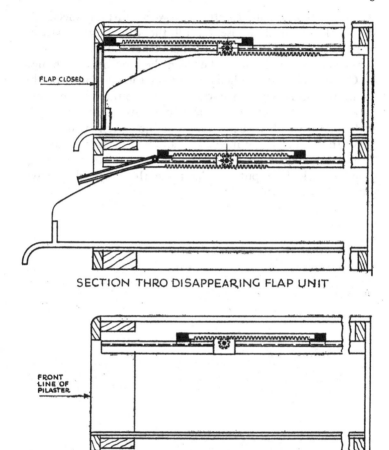

FLAP CLOSED

SECTION THRO DISAPPEARING FLAP UNIT

FRONT
LINE OF
PILASTER

SECTION THRO COMPARTMENT
SHOWING TRAY REMOVED.

When the tray is drawn out, the flap slides back and over. The mechanism is shown in the sections above and the sectional plan on the opposite page. This unit is designed by E. C. Stewart, as the result of the proceedings of the design research committee set up by Harris & Sheldon Ltd., described on page 66.

(Reproduced by courtesy of Harris & Sheldon Ltd., who hold a provisional patent for the design.)

the exchange of information between formal meetings; she is the clearing house for all the drawn and written work of the committee, and it is her responsibility to arrange contacts and appointments between the members. Although in Chapter IV, I touched briefly on the functions of the committee secretary, I would re-emphasise their importance here. The machinery of a committee must work smoothly and unobtrusively; there must be no hitch which will delay the work of any member; nothing, in fact, that will furnish anybody with an *alibi* when, at a formal meeting, they are asked to put on the table the work they have undertaken to perform.

CHAPTER VIII

The Effect of New Materials on Design

SEVERAL years ago manipulated steel tubing provided an entirely new type of material for manufacturers. It has since become familiar to most people in the form of furniture; its novelty has faded; but when steel tubing became available, it was not regarded by the furniture manufacturing trade as a material that promised new and stimulating economic possibilities. It was largely ignored; timber was the material the industry understood, they were organised to cope with it, and since the middle years of the last century they had been using wood for mass-producing much the same forms of furniture that had in earlier ages been made by hand. Woodworking machinery was highly efficient, and the rapid conversion of timber for furniture making was so conveniently organised, so widely understood in the trade, that the idea of a partial change over to another material was generally resisted by the manufacturers to whom it occurred. The furniture trade was muscle-bound by its own machinery.

With few exceptions, furniture manufacturers had seldom used timber really aptly; they seldom allowed the machine to do its best with wood. They studied books on antique furniture, and there was always the old hand-made Chippendale this, or Sheraton that, to misguide their direction of the superb mechanical tools engineers had given them. So it is not surprising that the possibilities of a new material such as steel tubing should have been appreciated and exploited by firms outside the furniture industry. Early in the nineteen thirties I was consulted by such a firm and together we were able to organise preliminary design

research work which led to successful results. Many enterprising firms in the Midlands, securing the services of able
designers, produced a number of light and occasionally
agreeable designs in steel and fabric and leather, and
achieved large sales for things like stacking chairs and café
tables in markets which had been the preserve of the furniture industry for generations. Precisely the same thing
happened when bent plywood furniture was developed in
Finland and introduced to this country. Bent plywood
invaded another section of the furniture trade's market.
In the same way, enterprising businesses concerned with
the manipulation of plastics, and perhaps with light alloys
too, may cut into the markets of trades that can think only
in terms of their traditional materials.

Apart altogether from such invasions of *old* markets with
new materials, the work of the design research committees
with which I have been associated in the field of plastics,
suggests that all kinds of hitherto unsuspected possibilities
may be explored. For example, hand luggage could be
transformed into something light, gay, flexible, unbe-
lievably expandable and convenient. All kinds of domestic
equipment could gain new shapes, colours and textures.
Much preliminary exploration has already been done by
design committees in such directions; though obviously,
under war conditions, materials could not be released for
experimental purposes, and designs could only be carried
to the preliminary model stage. But even so, much of the
initial thinking has been done, and by such far-sighted
research work in design, the time-lag between war and
peace production may be reduced.

Human progress has been recorded by the materials
which men have used for their tools, weapons and utensils.
From the Stone Age we passed to the Bronze Age, thence
to the Iron Age; we are now entering an age of Light
Metals and Synthetic Materials, of magnesium, aluminium
and its alloys and plastics. After the first world war the

great rayon industry was born. The influence of that industry has been world-wide. During the second world war the plastics industry has grown up: its revolutionary achievements are of comparable significance. There is hardly a large-scale industry which does not now use plastics in some form or other. If all the new materials, and the partnerships with old materials which they make possible, are valued for their intrinsic virtues and are not thought of merely as substitutes, we shall become realists about design, and be ready to employ their properties to the best advantage in holding, winning and extending markets for British goods.

I said in Chapter IV that the design members of a research committee often work for a preliminary period in a purely experimental way in order to become familiar with materials. This exploratory period is essential when they are dealing with such perplexingly varied materials as plastics and aluminium and its alloys. During that period the design members of the committee by working closely with the sales executive members may suggest ideas which, translated into terms of design, would enable manufacturers to improve their bargaining powers with producers of a material. It is obvious that an original design that affords a continuous and steady outlet for a group of materials, may affect the price at which the manufacturer can purchase his supplies. This economic aspect of design research work would be apparent in cases where the mass production of objects in certain types of plastics is planned. The whole question of price and the use of new materials is basically concerned with design.

In a paper on "The Influence of Plastics on Design," which I read before the Royal Society of Arts,* I briefly explored that subject, and also suggested the danger of making wild and foolish experiments with plastics, merely because those materials have such a potent appeal to the

* May 26, 1943.

imagination. There is a tendency to believe that plastics can do everything, that they are going to be the only materials to be used in the future, and that every kind of industrial problem may be solved by using them.

"There is a considerable danger in this state of mind, though perhaps it is not so much a state of mind as a state of intoxication. It may lead to the widespread misuse of plastics in industry, and it may encourage many people, after the war, to invest money in mushroom enterprises, which, by their ineptitude and frequent collapse, may discredit the new materials and hamper progressive development by reputable firms.

"There is another danger arising from intemperate enthusiasm for plastics; it is the assumption that their extensive use will automatically make familiar materials a back number. Some ill-informed writers have committed themselves to the statement that plastics will largely replace glass; which is about as sensible as saying that trousers will replace coats, or chairs will replace sideboards. Statements of this kind could be made only by people who still persist in thinking that plastics are substitutes: people without either the wit or the courage to understand that they are confronted by new materials. Again, such people imply that light alloys will be replaced by plastics, a statement manifestly unbalanced, and as basically unreasonable as saying that cheese will replace bread, or bacon will replace eggs. It is obvious to any student of the technique of industrial design that many materials have complementary uses, and that productive partnerships between materials are not only possible, but almost inevitable. I can imagine many such partnerships arising between various plastics and aluminium and its alloys, between plywood and plastics, between cast iron and plastics. Wasteful competition between materials in post-war industry can be avoided, particularly if the economic significance of industrial design is properly appreciated.

"In conducting research work in industrial design, I am occasionally asked: Are plastics going to be cheaper than —this, that or the other material? I don't know the answer to such 'questions: very few people would venture even on a rough guess. But the question shows a fundamental lack of appreciation of the potency of good design as a sales factor in the production of commodities. Good design forms a broad bridge between raw materials and consumer needs, capable of carrying an enormous press of traffic; and the excellence of the bridge increases the volume of the traffic.

"Nearly two years ago, in June 1941, the technical magazine, *Plastics*, published an unsigned article entitled: 'Prime Cost Bows to Design.' Under the heading of 'Cost and Value Relationships' the following paragraph appeared:

" 'Generally speaking, we insist that a new design, a new idea and greater stability of purpose are fundamentally more important than raw material cost. The mentality that utilises the formula, "What's the price?" first, conveys, to us at least, the impression of a mentality devoid of ideas and too ready to work down to a price, one which was the cause of price-cutting wars of the past, and eventually resulted in the use of cheap and poor raw material and skimped workmanship. Without purporting to be moralists our whole structure of production must be built on the idea that we should make a thing better than the other fellow, rather than that we must make it cheaper. If the better thing is as cheap or cheaper, so much to the good, but the two properties are not indissolubly hinged one to the other.'

"After the war, when plastics are released from their service to the war effort, we shall have abundant opportunities to prove that we can make things better than the other fellow; but unless research work in design has been undertaken, unless the native genius of British industrial designers is productively engaged, the influence of plastics upon the form and colour of thousands of things may be deplorable."*

It is not my purpose here to discuss the aesthetic aspects of new materials such as plastics, but to indicate their variety, their characteristics and the consequent need for adequate design research work if British industry is to make the best practical use of them. The recent growth and present range of the plastics industry may be briefly described.

Plastics are chemically produced materials which possess plasticity, and can be shaped by the application of heat and pressure. Only since 1935 has the term *plastics* become generally familiar in Britain. Although such materials had been produced and fabricated by British industry for over seventy years, not until that year was their contribution to contemporary life or their future possibilities widely appreciated.

In 1935 an exhibition of British Art in Industry was promoted by the Royal Society of Arts in collaboration with the Royal Academy, and held at Burlington House, London. The exhibit that set everybody talking and thinking, was devoted to the display of some astonishing materials which were described under the general name of "Plastics." The design of this exhibit was the work of a distinguished architect and industrial designer, Mr. Grey Wornum, F.R.I.B.A. He made the most of a unique opportunity. Seldom had any collection of strange substances been presented to the public with such vividness. Although people had become familiar with materials like "Bakelite" and such branded wares as "Beatl," this revelation of the prolific nature of the plastics family was staggering. That room in Burlington House brought into focus a picture of an unexplored and curious world. Even then, thinking people realised that a new type of industrial revolution was in the

* *The Journal of the Royal Society of Arts.* Vol. XCI, No. 4644, July 23, 1943, pp. 464-465.

making, though few would have ventured to predict what form it would take, or in what manner its growth would be accelerated.

The star exhibit in the plastics room was a piece of transparent material—utterly unlike glass, though equally transparent. Light sank through it, and gave it an odd depth of transparency. This substance was labelled "Resin M." It was described as a synthetic resin, and from the day of its first public appearance at Burlington House, the possibilities of plastics caught the imagination, not only of the British public, but of British industry. Thereafter, the dramatic character of plastics, their apparently unlimited forms, colours and textures, their astounding range of properties, were constantly referred to by popular writers; and in technical publications increasing attention was given to the effect of their use in many industries.

The ease with which they could be fabricated and their variety of colour brought them at first into the substitute class of materials. They were used as a convenient and economic way of gaining an effect that had formerly been achieved with wood or metal, stone, plaster or paint. Their true virtues were frequently ignored. Even when it was found that they could not only perform the tasks of older materials and do their work cheaply and well, but could contribute some entirely new services and qualities of their own, many manufacturers still felt obliged to make plastics simulate the character of the materials they had replaced. But in the development of new uses for these synthetic substances, the directors of industry were not wholly dominated by the idea of substitution.

The vast interest in the potentialities of plastics that was acquired by responsible people in British industry in the years that immediately preceded the second world war, had a profound and beneficial effect upon the British war effort. As I have said, plastics had been made in Britain for some seventy years, and had been used for all kinds of familiar objects—combs, knife handles, table tennis balls, toothbrush handles, telephones, car dashboards, buttons, switch plates, door handles, zip fasteners, dentures, to name only a few. The war accelerated the development of the industry. Hundreds of new uses were found; research and experiment were encouraged; thousands of problems connected with aircraft production were urgently needing rapid solution; supplies of most normal and accepted materials were diminished or interrupted, or could not meet the expanded demands of wartime industry. So plastics came into the picture—first (as in peacetime industry) chiefly as substitutes for familiar materials, and then, as invention and development swept along, they became materials in their own right. Achievements that would have been considered impossible in 1939, methods unthought of when the war began, substances and combinations of substances that were only in the theoretical stage ten years ago, became, by the third year of the war, almost commonplace.

The research chemist and the production engineer went into partnership, as metallurgists and engineers had gone into partnership over a hundred years earlier. That century-old association of technicians had produced a new world of steel, which altered the technique of shipbuilding and architecture. To-day, the chemist and the production engineer have brought nearer a world of new weights and measures, new ways of withstanding stresses and strains, new resistances to wear, tear and changes of temperature, new tensions. The vigour and development of the plastics industry in Britain, the enterprising and progressive outlook of the men who are directing it, and the inventive fecundity of its technicians, suggest that we may look forward to changes in the familiar shapes of innumerable things. New qualities, such as translucency and transparency, will be conferred upon the objects of everyday life, and on industrial products. Such anticipations are not mere flights of imagination; some hard-working reality of practical

THE Ekco radio set designed by Wells Coates, F.R.I.B.A.
This was first produced in 1935-6, and represented an
entirely new approach to the problem of designing a
compact and convenient cabinet in moulded plastics. The design
was influenced by the material, which gave the designer new
controls and new freedoms: the opportunity given by plastics
was thoroughly understood and used with originality.

*(The drawing is made from a photograph and is reproduced by courtesy of
E. K. Cole Radio Ltd.)*

achievement already sustains them, for the British plastics industry in wartime has proved that these vastly varied materials can stand up to tasks so exacting that normally only hard metals were supposed to be endowed with the endurance demanded for their effective and safe performance. Plastics are used for aircraft windows and cockpit covers, tank windows, aircraft stowages, eyeshields, floating torches, emergency ration packs, map cases; and some of the latest types are replacing rubber for insulation purposes. That is only the beginning of a huge list of uses. Even a short list of the plastics in general use would include at least twelve separate and quite different types. Such a list would be but a brief extract from the complete range of existing plastics. Within each different type there are numerous variations, and one and the same plastic may be opaque or transparent, harder than teak or oak, or of an elastic, rubber-like consistency, according to the way in which its basic constituents are combined.

All plastics may be divided into two main groups: *Thermo-Plastic* and *Thermo-Setting*. There is a third group which is based on the *Protein Caseins*. Thermo-Plastic materials are capable, like all plastics, of being moulded under heat and pressure, but do not change chemically and can be re-heated and re-formed. Scrap can be used, so there is no waste. Thermo-Setting plastics, when once they have been shaped by heat and pressure, remain hard and unchanging, and can only be altered by cutting or grinding. Each of these two main groups include a number of types which conform to the basic character of their group, but differ from each other in quality and in the sources from which their raw materials are derived.

The Thermo-Setting group includes Phenolic resins, Cast Phenolic and Urea resins. In the Thermo-Plastic group are Cellulose Nitrate, Cellulose Acetate, Acrylic resins, and Polystyrene. The minor group, Protein-Plastic, is derived from Casein.

In the *Thermo-Setting group* are the following:—

Phenolic resins. They have resistance to shock, heat, acid and alkali. They are free flowing for extrusion or transfer moulding. Their chief use is in compression moulding.

Cast Phenolic. This has high impact strength, rigidity, is non-inflammable, and is easily coloured and fabricated.

Urea resins. They are translucent and have a large colour range; they are splinter-proof, light in weight and rigid. They have a hard surface and they diffuse light well. They are almost without taste or smell.

The *Thermo-Plastic group* includes the following:—

Cellulose Nitrate. It is tough, but easy to fabricate, it is also easy to cement and to colour. It resists the action of water and possesses real transparency.

Cellulose Acetate. It is tough, light, alkali resisting, odourless, and can be easily fabricated and coloured.

Acrylic resins. They have rigidity and lightness, they resist water and weather, and take colour easily. They have exceptional transparency.

Polystyrene. This gives a great range of colours, and can be either transparent or opaque. It has a high index of refraction; it can transmit light, and it resists chemical action.

Finally, there is the *Protein-Plastic group*, based on Casein. Plastics in this group are non-inflammable, easy to colour, polish and fabricate.

Although the plastics industry originated in the middle years of the last century, its present form is of comparatively recent growth. For instance, celluloid is such an accepted and familiar material that it is not always recognised as a plastic. But it comes into the Thermo-Plastic group, and is actually the oldest plastic to be used commercially and developed beyond the experimental stage, though to-day it has to face keen competition from other more versatile and non-inflammable types. The first patent bearing on celluloid was taken out by a Birmingham scientist called

F

Alexander Parkes, in 1855. But the real beginning of the modern plastic industry dates from the patent taken out in 1909 by Dr. Leo Baekeland for the plastic known as "Bakelite." It is one of the most widely used plastics; the tonnage of its production is probably the greatest for any one type. There are many slightly varying plastics made from the basic combination of Phenol and Formaldehyde by the large plastics manufacturers of Great Britain, though only one has the right to be called "Bakelite," a registered name which is the property of the Bakelite Company. In the last fifteen years many other new types of plastic have been produced, and during the second world war some spectacular new materials have appeared. Wartime production has enabled so many ideas to be tested out in conditions far more severe than any peacetime use could impose, that the British plastics industry is already far ahead of its 1939 achievements. It has packed into the war years the experience and development, experiment and research, that would normally occupy a decade. But without corresponding research in design by industries concerned with fabricating plastics much of this experience may be wasted or unprofitably employed.

The economic significance of such design research is underlined by the anonymous authors of that informative book, *Plastics in Industry*. The final paragraph of their last chapter reads thus:

"It has been suggested that what is really needed to-day by the plastics industry is a central bureau of designers properly trained and thoroughly well fitted to deal with all types of commissions and problems concerning plastics. This idea will not, we feel sure, be acceptable to the small moulder who, although quite aware of its possible advantages, cannot afford, or thinks he cannot afford, any design cost whatsoever. As stated earlier . . . his normal process, if he has no designer, is to approach his mould-maker who is usually completely devoid of any real artistic ability and

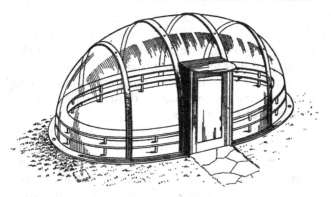

SOME POSSIBLE USES FOR PLASTICS :

In these tentative designs for small green-houses, it should be realised that the capabilities of the material have guided the ideas of the designer. Greenhouses of this type and with these variations of shape, would be suitable for small gardens and the materials from which they could be made already exist and have given good service in war industries.

It is suggested that the curved supports are made from flat transparent extrusions, bolted to an outer rim of coloured opaque extruded plastic. The door supports, frame and canopy are of laminated opaque sheets cut to shape and the door panel is of transparent plastic. The covering is of transparent formed Cellulose Acetate or Acrylic sheet. Shelves and supports are of laminated opaque sheets and rods, cut to shape. The outer rim is bolted down in small concrete beds. The variations include a shape designed for a restricted space, to go against a fence in a sunny position.

(Designer: Harry Jones, N.R.D..)

is only capable of turning out an engineering job. When the moulder attains his object by getting the tool-maker to adapt an existing design he is under the impression that he does not pay for it as his invoice usually makes no mention of this service. This is, of course, a gross error, as the time spent on the work by the toolmaker's draughtsman is added on to the total cost of the tools."*

Everything has to be paid for, whether the cost is apparent or concealed, and lack of research in design sometimes costs a market.

Preliminary design research work, carried out to demonstrate the capabilities of some particular material or group of materials, may be undertaken merely to instruct industrial designers, as part of the short educational period I described in Chapter IV. But from exploratory studies by imaginative designers of the properties of such materials as plastics and light metals, may arise new ways of approaching markets with new products. Such new products may be astonishingly novel, when the properties of materials like aluminium and its alloys are adroitly used to solve some familiar problem. The significance of this type of experimental research in design is demonstrated by two examples, concerned with these light metals, of which I am able to give specific details.

A few years ago Aluminium Union Limited invited various industrial designers to choose some particular problem that could be effectively solved with aluminium, and to do some experimental designing. The chief properties of aluminium and its alloys may be briefly summarised as follows: they are ductile and easily and economically cast and fabricated. Perhaps their most outstanding characteristics are high strength to weight ratio and resistance to corrosion. The designers, after preliminary discussion in which they selected their problems, worked with technical experts in the material. Among the problems they tackled and solved were a portable

* *Plastics in Industry*, by "Plastes" (Chapman & Hall Ltd., 1940), Chapter XVII, p. 234.

electric fire and a trailer caravan which could be expanded
so that it became a week-end house. Those two experimental designs were produced by Mr. Christian Barman
and Mr. R. D. Russell.

Mr. Barman designed the portable electric fire, and he
selected this particular problem because the material had a
sequence of properties which allowed a designer to depart
from the methods that had hitherto been accepted for the
form of electrical heating appliances. Electric fires have
nearly always retained their kinship to the open fire; except
in one or two designs, they suggest that some glowing
embers or red hot bars are framed, as a coloured photograph
of a fire might be framed, in a panel on the wall or in some
movable setting, either of metal or some kind of fire-resisting
composition. Mr. Barman set out to design a portable fire
which would never be in the way, which could be put down
anywhere, and could be overturned without danger. He
produced a featherweight spherical cage or basket, formed
by curved vertical and horizontal bars of anodised aluminium in contrasting colours, with a base of spun sheet
aluminium, and a ring-shaped handle of some transparent
plastic, such as cellulose acetate. The heat element was a
rod placed vertically in the centre of the basket, and this
basket could be rolled about the floor and nothing that
came into contact with it was ever in danger of catching fire.

A very different problem on a vastly different scale was
the designing of a trailer caravan. Mr. R. D. Russell fully
realised that the lightness of aluminium and its alloys
creates the most revolutionary possibilities in transport.
Aluminium is only one-third the weight of other materials
with the same tensile strength. By using such materials, an
entirely new form of caravan, light enough for easy towing,
became a practical possibility. It was designed to comply
with legal requirements under the Road Transport Acts,
with a maximum overall width of 6 feet, and it could be
expanded on a temporary site to the dimensions of a small

A PORTABLE ELECTRIC FIRE

designed by Christian Barman, F.R.I.B.A., N.R.D.

This is a radiant fire, and the high reflectivity of the principal material is essential: aluminium will reflect up to 85 per cent of average radiation—only pure silver will do more. There is secondary heating by convection. The fins at the back of the cast aluminium reflector quickly reach the high temperature required. As a portable fire, it must be light; and the whole design, including a length of aluminium flex, weighs only a few ounces. Light objects are easily upset, but the spherical basket-guard allows this fire to be overturned without danger.

ELEVATION OF FIRE

The basket is built up from vertical and horizontal rings. The vertical rings are of heavy gauge aluminium wire. The horizontal rings are flat sections cut from sheet metal.

PLAN

The rods and flat rings are of anodised aluminium of contrasting colours. The ring-shaped handle at the top is of plastic.

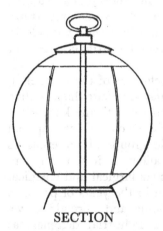

SECTION

The heat element is a rod placed vertically in the centre of the basket. The polished aluminium reflector is concave in the middle, convex at the sides.

The heat beam from the reflector has a wide lateral spread.

In the vertical plane the beam is more highly concentrated.

The air in the room is heated by an arrangement of convector fins.

(Reproduced by courtesy of Aluminium Union Limited)

week-end house. The high strength to weight ratio of aluminium and its alloys, and their insulating properties, largely controlled the design of this trailer week-end house. It had to be light enough to be pulled by a small car and strong enough, when extended, to be quite rigid; it had to be reasonably cool in hot weather and easy to heat in cold. Also the material had the advantage of being unpopular with earwigs, ants, and other insects which frequently over-crowd a caravan or tent in summer. In construction the manufacturing techniques used in aircraft production were employed. The walls, roof and floor were slabs of stressed-skin type consisting of aluminium sheets fixed to braced framing with a packing of alumimium foil to provide insulation.

The trailer was designed to be used either closed for short periods—a night or two during travelling—when four people could sleep in it as comfortably as in a normal caravan, or open for long periods, the latter being far the more important. Thus, all storage and essential equipment was available in either state, being in the rigid part of the structure and not covered by moving parts.

With such practical examples before us of the manner in which the properties of materials may stimulate the inventive powers of trained industrial designers; with the knowledge that the second world war has brought into special prominence two such remarkable groups of materials as plastics and aluminium and its alloys, is it too much to claim that the war-born gifts of metallurgical and chemical research could be used with vision in the years of peace to re-establish British industry at home and abroad? If we have vision, we shall employ our industrial designers as well as our other technicians. We could use our vast latent capacity to produce well-designed goods which were within the economic reach of the people and were attractive to foreign markets.

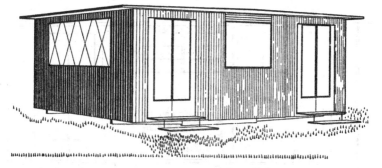

A TRAILER WEEK-END HOUSE

Designed by R. D. RUSSELL, N.R.D. Above, the trailer is shown extended as a week-end house: ·plans and sections of it, closed and open, are given on the next two pages.

To be extended, the trailer must be drawn on a fairly level site. Two telescopic cantilever girders, which are hinged in their length to hold up horizontally during transit, are let down to their vertical positions. On the underside of these girders is a hydraulically actuated self-levelling jacking system which is then put into operation so that the centre of the trailer is supported off the wheels and is level in both directions. The roof extensions are hinged along their top edges to the fixed centre roof and are lifted into position, being supported by folding stays. The telescopic girders are engaged in slots in the bottom of the main extending side walls (shown blacked in the details on pages 90 and 91) and a double-acting worm gear housed inside the girders is operated to extend them. The extending girders push out the main side walls—these are strengthened at the bottom by wide angle members to take the thrust—and the main side walls pull open the triple-hinged wall sections at back and front. During the whole operation the weight of the extension walls is carried by the cantilever girders. The roof extensions are then slightly lowered on to the walls and engage to form a tight joints The floor extensions, which are hinged along their lower edges to the fixed centre floor, are then lowered into their horizontal

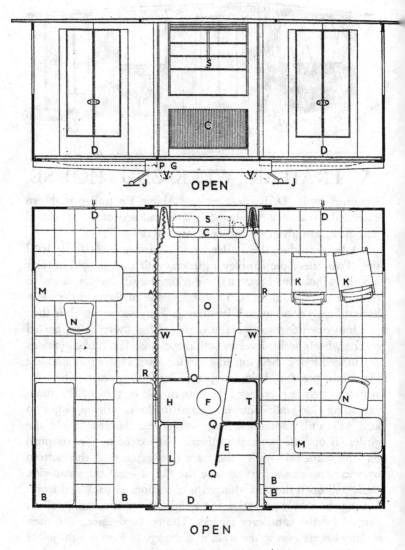

OPEN

OPEN

(Reproduced by courtesy of Aluminium Union Limited)

KEY TO SECTIONS (*above*)
AND PLANS (*below*)

A. Curtains either shutting off whole extensions as bedrooms or enclosing only during day.

B. Bed. The frame is an aluminium tube on to which spring strips are mounted forming a shallow basket. During transit or when used as bunks, the moulded rubber mattresses fit inside these baskets, as beds, they are laid on top of the inverted frames.

C. Cupboard with tambour front containing calor gas cylinder, gas refrigerator and storage space. Cooking units and a small sink are in the top.

D. Doors to outside.

E. Chemical closet. The closed container is removable from outside.

F. Roof light.

G. Telescopic girder shown folded up in closed section and down in open section.

H. Hanging space for clothes.

J. Hydraulically actuated self-levelling jacking system.

K. Folding easy chair.

L. Ship type hinged lavatory basin with storage cupboards over and beside.

M. Folding tables with tops hinged in two sections to stack in space beside wheel runs.

N. Nesting chair.

O. Compressed cork tiling.

P. Increased thickness in centre part of floor extension to give necessary strength.

Q. Sliding doors.

R. Curtain track built in flush with ceiling.

S. Glazed cupboard for cooking and eating equipment. The cupboard is backed by a window.

T. Trays for clothes.

V. Wheels.

W. Wheel runs with storage space beside.

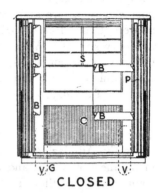

CLOSED

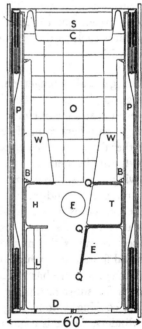

60″

CLOSED

position and engage with the bottom of the extension walls. The floor extensions are also carried on the cantilever girders and, between these girders, their thickness is increased to give the necessary strength. Where the large window openings occur in the main extension walls, the structural bracing members are taken uninterrupted through these openings to maintain the strength of the slab.

The outside sheets of the walls generally are of corrugated aluminium of about one inch pitch, the corrugations running vertically. All glazing is in transparent plastic sheet for the sake of lightness of weight and elasticity. The floor is covered with compressed cork tiles. The metal is anodised throughout as a protective treatment. The corrugated outside surfaces are clear olive green. The roof, including edge and soffit, all outside plain surfaces such as doors and window frames, and all metal inside are white.

The general arrangement of the plan is self-explanatory. (See page 90.) When extended it divides roughly into areas for cooking, washing, storage of clothes, sitting, writing and sleeping. The sitting area is kept rather free in order to give a sense of spaciousness. Either the whole of each extension, which forms a bedroom, or the beds only in their recesses can be shut off by curtains, the former giving a reasonable amount of privacy for two families or for parents and children, the latter making it necessary to tidy up the bedrooms before breakfast. The beds can be used as a fixed seat or as bunks—they are the latter in the closed trailer. When used as bunks the moulded rubber mattresses go inside strip-metal foundations which are in the form of shallow open baskets. All the furniture has frames of aluminium and either folds up or nests for stowage during transit or when not needed. Lighting and cooking is done by gas stored as a liquid under pressure in cylinders. There is a gas refrigerator in the cupboard below the cooking and sink units.

CHAPTER IX

Old Materials with New Properties

IN the last chapter I have emphasized the possibilities of
plastics and light metals, but the glamour of those materials
should not be allowed to obscure the fact that there are
many others, long-established in use, which have, as the
result of creative research work conducted by their producers,
acquired new properties. When such new properties are
studied by industrial designers in collaboration with producers, the creative work that began in the laboratory is
continued, and fresh applications and new techniques
generally follow. This logical sequence of creative operations
can invigorate not only the activities, but the collective
thinking of an entire industry. For instance, the manufacture
of structural and decorative glass has for many years been
conducted with an alert regard for research, with an understanding of the function of design, and with a discerning
capacity for relating such complimentary operations to
the character of existing and potential markets. The
diversity of British glass manufacturers' achievements
was recently indicated in a paper read before the Royal
Society of Arts by Professor W. E. S. Turner.* He referred
to a development which may bring about many changes in
the design of shop-fronts and the manner in which goods
are displayed in shop windows. This new form of shop-
window glazing "is to substitute ordinary plate glass with
its faint greenish tint by plate having a slight tint to set off
the colour of goods displayed. These special glasses have

* *New Uses for Glass*, by Professor W. E. S. Turner, O.B.E., D.Sc.,
F.R.S., of the Department of Glass Technology, University of Sheffield,
Journal of the Royal Society of Arts, Vol. XCI, No. 4636, April, 1943.

been developed by a large glass manufacturing firm in this country under the name of 'Spectralite.' "* Professor Turner mentioned some of the partnerships between materials that have within recent years furnished manufacturers with opportunities. "Combined with plastics, glass fibre has been converted into fireproof structural material with wide applications." He also pointed out that "great advances have been made in the understanding and technique of sealing glass to ceramic materials and metals."†

Perhaps one of the most revolutionary results of war-time research is the welding of aluminium directly to glass. In another book I have summarised a few of the comparatively recent advances in the production of specialised glasses. For example: "There is a glass that admits a high proportion of natural ultra-violet radiation, and these powers of admission are permanent. There are also glasses of orange and bluish-green hues, which are used in jam factories, abattoirs, dairies and in some larders, which cure the fly nuisance in summer. Flies dislike the special tints which these glasses impart to daylight.

"A multitude of decorative glasses exist, such as rectangular units of mirror mounted on fabric, which can be bent round columns and which can curve this way and that; and opaque, coloured glasses, and patterned glasses of various kinds. Opaque glass in ashlar sizes, or in large sheets, makes an admirable external skin for a building, for, like tiling, it can be kept clean by hosing. . . . There are glasses which can bend light, and turn it into dark corners. These are prismatic glasses which can direct light to the depths of the gloomiest passage, and give a generous share of natural illumination to basements and ground-floor rooms that would normally be dark."‡

* *New Uses for Glass*, p. 232.   † *Ibid.*, pp. 227–228.
‡ *The Place of Glass in Building*, edited by John Gloag (Allen & Unwin Ltd., 1943), p. 13.

There is also the remarkable toughening process which, applied to plate glass after manufacture, confers great powers of resistance to heat and shock and gives sufficient flexibility to allow the material to bend and twist under pressure. Glass neatly reinforced by wire netting of square mesh with the intersections of the wire electrically welded, has been available for several years. Such technical achievements support Professor Turner's conclusion that "the undue fear of glass as a brittle substance is slowly passing away."

Other familiar materials have regained some of their original novelty as the results of fresh research reveal new properties, new finishes and new associations with various substances. In the late eighteenth century cast iron was as exciting as plastics are to-day. I have described and illustrated in Chapter I the first cast iron bridge at Coalbrookdale. In the biography of one of the greatest architects of that time, John Summerson says that "in those days people sometimes affected unreasonable enthusiasm about cast iron, just as they do to-day about reinforced concrete. Those who went to Coalbrookdale and saw what John Wilkinson and Abraham Darby were doing, came away full of ideas about the new material; and even if they did not, like Wilkinson, order a cast iron coffin and a cast iron mausoleum, they were not likely to miss any opportunity that occurred of trying out the capabilities of the material."*

The history of cast iron as a material is instructive. During the early decades of the nineteenth century it continued to excite the enthusiastic regard of architects and engineers. John Rennie designed many bridges in cast iron, though his proposal to span the Menai Straits was considered too daring. "Mr. Rennie proposed to accomplish this object by a single great arch of cast iron 450 feet in span—the height of its soffit or crown to be 150 feet above high water at

* *John Nash: Architect to King George IV*, by John Summerson (Allen & Unwin Ltd., 1935), Chap. II, pp. 43-44.

spring tides."* Apart from such large-scale uses, there were innumerable decorative ways in which cast iron was employed, ranging from railings and balustrades, to elaborate and delicate filigree castings for such domestic articles as fruit dishes. An immense amount of skill was available in foundries; some of the early mould-makers were superb craftsmen; but owing largely to the absence of the designer from industrial production, this skill was undirected. Cast iron was used to imitate ornamental forms that had been evolved with other materials; it became a cheap dodge; and it remained in this substitute phase for a long time, with the most damaging results to its reputation. Architects, who were once the most ardent advocates of the use of cast iron, are now inclined to associate the material with the bandstands, public conveniences, and lamp-standards, smothered with "Gothic" ornament, which were produced in the latter part of the nineteenth century when manufacturers were doing their own designing, and advancing year by year into a deeper and darker jungle of complicated decoration. To the public, cast iron became identified with the more repellent branches of housework; the black-leading of the kitchen range and the parlour grate was inevitably associated with a material that for three-quarters of a century had been used for coal-burning appliances without any consideration being given to labour-saving. The material was squandered on such domestic uses; stoves and grates were far too big, too complicated and too clumsy. Rudyard Kipling, in a story about an American inventor, put into the mouth of this character a penetrating comment: "The British think weight's strength."† Coal-burning appliance makers certainly supported this criticism in Victorian and Edwardian times by their ponderously lavish use of cast iron: apparently they wanted grates and ranges to look

* *The Lives of the Engineers*, by Samuel Smiles (John Murray, 1862), Vol. II, Chap. VI, p. 175.
† "The Captive," in *Traffics and Discoveries*.

The G.P.O. public telephone call box, designed by Sir Giles Gilbert Scott, PP.R.I.B.A. It is made from the most appropriate and enduring material, cast iron. The mass production of thousands of these boxes was facilitated by the choice of material and the designer made admirable use of its properties. (*The drawing is made from a photograph, and is reproduced by courtesy of the Postmaster General.*)

G

heavy: they succeeded: the results are still cumbering fire-
places and kitchens in thousands of homes. Although industrial designers have been employed by some appliance-
makers within recent years, and designers have occasionally
made use of cast iron in other directions, the mishandling
of the material between the middle years of the last century
and the nineteen-twenties has caused many people to regard
it as hopelessly old-fashioned, an untenable view that is
soon corrected by knowledge of the progressive improvements in the qualities and capacity of cast iron achieved
by contemporary research. How admirably it may be used
to-day has been demonstrated throughout the length and
breadth of the land by the telephone call-boxes designed
for the G.P.O. by Sir Giles Gilbert Scott, PP.R.I.B.A.
How versatile it has become is revealed by the work of the
British Cast Iron Research Association and by the research
programmes of enterprising individual manufacturing firms,
which have made available an array of new finishes and
new grades of the material. For example, the surface of cast
iron can be chemically changed to render it stable and
highly resistant to corrosion: permanent skins of other
materials, such as aluminium or zinc, can be sprayed on
at high temperatures. Metallic, enamel and organic paint
finishes have finally dispelled the black-leading bogey from
domestic appliances.

Cast iron is a material derived from a primary process:
it is convenient, cheap and, compared with other materials
which can be cast, its strength is considerable. The properties of such grades as high-duty cast iron, are remarkable: they have been tested and found good by the stupen-
dous exactions of war service. It is one of the many established materials that may derive great benefits from research
in design, and when its attributes are intelligently and
imaginatively employed it may solve many problems of
industrial design in the future.

Although in this and the previous chapter I have used

only four materials as my principal examples, I believe that there is an economic and useful place for every material, whether it is mined, grown or chemically produced; but that belief can be sustained only if British industry realises that while materials have an effect upon design, the operation of design may have the most stimulating effect upon the character, use and future of materials.

CHAPTER X

National Character in Industrial Design

URING the twenty years between the two world wars we had to fight hard to keep our foreign markets, and sometimes we used old-fashioned weapons. We had become so accustomed to "sound British workmanship" selling the goods for us, that although we were no longer the "Workshop of the World," many responsible people seemed unable to appreciate that we had lost the title. Very few of the men who had fought in the first world war were in positions of control, during the 'twenties. In those days, some firms still approached overseas markets with such casual arrogance that they never even thought of translating their catalogues into the language of the country where they hoped to sell their goods. Although the "take or leave it" attitude to selling may have worked in mid-Victorian times, as a technique of trading it was never calculated to endear us to other nations.

Our apparent reluctance to make any fundamental changes in the character of the goods we supplied to markets abroad and at home became the subject of fierce and continuous criticism by writers, designers and educational bodies whose activities were dedicated to propaganda for the improvement of design in industry. Much of this criticism came from people who had no knowledge of industry, who were prejudiced against business, and were often confused in their own minds about the nature of industrial design. There were some important exceptions: perhaps the most active of the various propagandist bodies in the 'twenties was the Design and Industries Association, which included among its officers such men as Frank Pick and Sir Lawrence

Weaver, great business executives who understood the practical significance of design. Frank Pick's work for London's Underground Railways, trams and 'buses, was one of the most outstanding examples in the world of the large-scale operation of industrial design.* But even the Design and Industries Association was confused in purpose; its members were inclined to muddle up the improvement of industrial design with the preservation of handicrafts. From 1922 to 1927 I edited the Year Books of the Association, and glancing through their pages to-day, I can see how often the illustrations selected by the editorial committee innocently complicated the whole issue. In those days, as I mentioned in Chapter III, I was inclined to believe, in common with many members of the D.I.A., that "the designer is always right." My subsequent practical experience has led me to qualify that belief, so I would now say: "the *appropriate* designer is generally right."

The appropriate designer for industry is seldom the man who believes so passionately in the preservation of handicrafts that he despises machinery. I am not asserting that interest in handicrafts is invariably incompatible with interest in industrial design; but I do suggest that they are separate interests, usually demanding separate outlets for those who seek creative expression in terms of design. It was the confusion of these interests that invalidated much of the criticism launched at the British manufacturer by such bodies as the Design and Industries Association. The D.I.A. had adopted as its slogan "Fitness for Purpose," and an excellent slogan it was in the 'teens of the century, when the Association was founded, for at that time innumerable mass-produced articles were just shoddy accumulations of ornament. It was a campaign cry for cleaning-up such rubbish; but it was not an end in itself.

Throughout this book I have deliberately avoided the

* I have attempted to describe the scope of Frank Pick's work in Chapter VIII of my book *What About Business?*

discussion of aesthetic or educational questions concerned with design; but in considering the expression of national character in industrial design, and its particular significance in overseas markets, it is impossible to ignore the volume and nature of the criticism to which British goods have been subjected between 1919 and 1939. It is a national habit to label any departure from traditional ideas in design with some slick name, often borrowed from some movement in art that has won acceptance as a fashion. For instance, almost any unusual shape or unfamiliar decorative pattern was, up to a few years ago, called "Cubist" or "Futurist"— with or without abusive emphasis. The language varies from decade to decade. Late in the 'twenties the word "modernist" was first bandied about: soon it degenerated into "modernistic." Then, in the early 'thirties, we began to hear more and more about "functionalism," which was the highly dramatised rediscovery of common sense in design, and was, indeed, the practical expression of the slogan "Fitness for Purpose." I have no wish to belittle the virtues of "functionalism" by challenging its novelty, but as a guiding principle it has a respectable ancestry. The architect Vitruvius, writing in the time of Augustus, said: "In architecture, as in other arts, two considerations must be kept constantly in view: namely the intention, and the matter used to express that intention; but the intention is founded on a conviction that the matter wrought will fully suit the purpose. . . ."*

The latest label is "streamlining," which is a technical term in aircraft design. The word "streamline" also denotes the natural course of water or air currents. The term has now become accepted as a popular description for the smoothing over and tidying-up process that is inseparable from the design of modern locomotive machines, aeroplanes, cars and railway engines. The growth and progress of this

* *The Architecture of Marcus Vitruvius Pollio*, translated by Joseph Gwilt (1826), Book I, Chap. I.

streamlining process is indicated in some of the plates at the end of the book; and it has affected the form of many mass-produced articles. Actually it is an exceedingly old technique of design: in A.D. 1000 it is recorded in the Saga of King Olaf Trygvesson, that Thorberg, the builder of the ship *Long Serpent*, made deep notches in the hull which he afterwards smoothed out, thus improving the shape of the vessel by giving it sleek, graceful lines.* It is also recorded that King Olaf, whose views on shipbuilding were traditional, threatened Thorberg with execution, until the finished job convinced him that a new and better type of hull had been designed.

All these various labels, such as "functionalism" and "streamlining," which indicate a point of view or a fashion or a genuine improvement in the technique of design, are apt to make people with progressive minds, particularly designers, forget or ignore the importance of *nationality* in design. Raymond Loewy in his paper, *Selling Through Design*, which I have already quoted in Chapter V, states that "it has become apparent that design in order to remain useful commercially, must be universal aesthetically. Progressive English designers admit that the domestic quality of some of their designs has closed many markets to English manufacture. The quality of English manufacture is unsurpassed; no designer has the right to handicap a product by restricting its sale in foreign markets simply because an international character is lacking."†

While many things designed expressly for the home market might be unsuitable for foreign markets, this rather obvious fact does not support Mr. Loewy's view that design must be "universal aesthetically" if it is to be a commercial proposition. He seems to be making a plea for something so dull and deadening that it is difficult to take him seriously, for those who have met Mr. Loewy know that nothing dull

* Chapter XCV, "The Building of the Ship Long Serpent."
† *Journal of the Royal Society of Arts*, No. 2604, Vol. XC, page 99.

could ever be lodged in his agile mind. So many designers who want to practise what is called the "international style" are exponents of that functionalism which Mr. Loewy condemns when he says "the cold, uninspiring examples of so-called Functionalism are rebukes to the profession of Industrial Designer."* When, after the second world war, we seek new markets abroad and attempt to expand those we already hold, *what* can we offer to potential consumers in Asia, Africa, South America and the Dominions of the Empire? Pure functionalism, unlit by imagination, will not impress them, nor would they accept, save on a price basis, goods designed in an "international style" which has ironed out all the little peculiar national and regional variations of taste and fancy and represents a uniform, standardised approach to almost every problem of industrial design and production.

By the expression of national taste in design, I do not mean the reproduction or adaptation of traditional ideas. For far too long "olde England" has been dished up for foreign consumption. For many years manufacturers of all manner of products have put over the idea that Queen Anne is not dead, but alive and implacably unprogressive, and that the first four Georges are still on their thrones. To rake up ideas from the past and to misuse our great new industrial capacity for reproducing them, would be a disastrously retrogressive policy—a form of commercial suicide. Our goods can speak our language clearly to the world without the mannerisms of former centuries; our modern accent will be recognised and acceptable, and we have something to say that is worth saying. Our own able industrial designers, working with manufacturers on the lines I have suggested in this book, or in any other way that allows freedom for experiment and research in design, could give to our goods the one quality that precludes competition: national character.

* *Journal of the Royal Society of Arts*, No. 2604, Vol. XC, page 94.

Old-fashioned private enterprise was often a game of beggar-my-neighbour: new-fashioned private enterprise can change that silly game to better-my-neighbour. We could better all our neighbours in matters of design.

We've got the skill: we've got the men: at home and abroad the markets await us.

Appendix

THE following are the addresses of the principal organisations concerned directly or indirectly with industrial design:

Central Institute of Art & Design,
National Gallery,
*LONDON, W.C.*2.

Design & Industries Association,
National Gallery,
*LONDON, W.C.*2.

National Council of Art & Industry,
National Gallery,
*LONDON, W.C.*2.

National Register of Industrial Art Designers,
National Gallery,
*LONDON, W.C.*2.

Royal Institute of British Architects,
66 Portland Place,
*LONDON, W.*1.

Royal Society of Arts,
10 John Adam Street,
*LONDON, W.C.*2.

Index

Introduction to the Plates

THE plates have been selected to show the operation of industrial design in a variety of fields, from such large-scale problems as the convenient disposal of equipment and apparatus on board a modern liner to the shaping of cooking glass. The evolutionary process of industrial design is revealed by a diversity of examples: locomotives, railway coaches, electric trams, and automatic ticket machines have been selected to illustrate progressive development.

PLATE I SUPERSTRUCTURE OF THE S.S. *ORCADES*, ORIENT LINE

Designer: Brian O'Rorke, f.r.i.b.a. This is a view of the upper decks. In consultation with the shipbuilders and naval architect the whole superstructure was "tidied up" by a reduction in the number of supports, grouping of scupper-pipes, etc., and a less haphazard spacing and arrangement of minor details. Even the most recent British ships present a maze of auxiliary equipment and unrelated excrescenses. This tidying up process gives increased comfort for the passenger, and is likely to increase operating efficiency.

(See page 47.) (*Reproduced by courtesy of the Orient Line.*)

PLATE II   *Above:* Throughout the latter part of the nineteenth century modern ideas of design gradually emerged: by the opening years of the present century, the clean, uncomplicated lines of locomotives were already foreshadowing what came to be known as the "Modern Movement." This Caledonian Railway express engine of the "Cardean" class, is a prophetic example. Designer: J. F. McIntosh.
(Reproduced by courtesy of the L.M.S. Railway.)

Centre: This Great Northern locomotive, of the type used for hauling the "Flying Scotsman," was typical of the advance of design in the nineteen-twenties. Designer: Sir Nigel Gresley.
(Reproduced by courtesy of the L.N.E. Railway.)

Below: The next stage of development in design. This locomotive represents one aspect of the "tidying up" process—the smoothing off of awkward corners and angles—that is everywhere apparent in the design of mobile machines. Designer: Sir Nigel Gresley.
(Reproduced by courtesy of the L.N.E. Railway.)

PLATE III Rolling stock on the District and Metropolitan Railways in London has been progressively improved in design.

Above: A District Railway car in 1905.

Below: Cars put into service late in 1937.

(*Reproduced by courtesy of the London Passenger Transport Board.*)

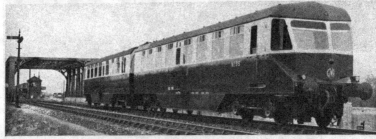

PLATE IV   Bigger and better windows characterise the design of modern railway rolling stock and road vehicles. The old stage-coach window prototype has gradually been abandoned.

Above: The Beaver Tail Observation Car on the L.N.E.R. "Coronation" Express. The curved window at the back is of transparent plastic.

Centre: An oil-engined rail-car operated by the Great Western Railway.
(Reproduced by courtesy of Associated Equipment Co., Ltd.)

Below: A motor coach that admits the maximum amount of light and air.
(Reproduced by courtesy of Associated Equipment Co., Ltd.)

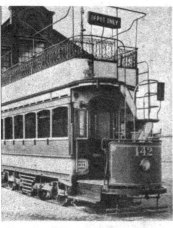

PLATE V   TRAM TIDYING

Above, left: A car on the London United Tramways, in service in the early days of the present century. The driver was pushed out in the cold and the rain.

be*low, left:* The covered-top tram still left the driver unprotected, and consisted of two long glazed boxes, one on top of the other.

ab*ove, right:* The tidying process has been carried out, and here is a tram with sleek bean lines, a cabin for the driver, and increased comfort for passengers.

be*low, right:* The tram goes off the lines. The trolleybus makes the best of both possible worlds—electric traction and the mobility derived from independent steering.

(Reproduced by courtesy of the London Passenger Transport Board.

PLATE VI *Above, left:* An early type of automatic ticket machine in service on stations of the London Underground Railways.

Below, left: The next phase of development in design.

Above, right: A compact battery of ticket-issuing machines with all superfluous ornament, projections, and odds and ends removed.

Below, right: The automatic booking office has arrived: the ticket machines are now sunk in a wall and make no claim on the floor space of a booking-hall.

(Reproduced by courtesy of the London Passenger Transport Board.)

PLATE VII   *Left:* Bus stops are marked by a sign carried on a post. This post may be of wood, cast iron, steel or concrete. This fine and delicately moulded cast iron post was produced by the London General Omnibus Company.

Centre: Bus time-tables were originally displayed on walls or fences near a stop. A better way was to display them on the stop post itself. The post was fitted with a well-made hardwood box with glass windows on three sides; but this involved maintenance and repairs. The wooden box, however well-made, was liable to warp and crack, so experiments were made with bronze boxes, but these were costly.

Right: The proper solution was to discard the original post design and start again at the beginning. Result: a one-piece post with a pair of fins in which the time-table frames are incorporated. By use of Terrazzo, maintenance is eliminated. Experiments were made with various mixtures, and a combination of white marble chippings and grey cement had the best weathering properties.

PLATE VIII

The "Ultimat" patent automatic whisky pourer and measure made from transparent and opaque plastics. The plastics employed are unaffected by the action of alcohol, and a clean, uncomplicated piece of design results from the apposite use of these materials. Designed for Gaskell & Chambers by British Industrial Plastics Ltd.

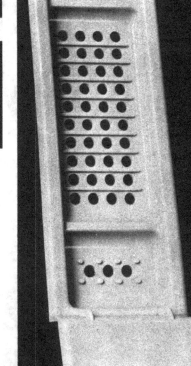

PLATE IX

A bath tray, a toilet roll holder and a tooth brush holder in ivory white opaque plastic. These plastic mouldings were designed by Leslie Mansfield and James Barnes.

(Reproduced by courtesy of Pilkington Brothers Limited.)

PLATE XI

COMBINED PELMET, CURTAIN RAIL AND VENTILATION UNIT.

The section is in extruded aluminium and combines the separate items of pelmet board and brackets, curtain-track and fixing brackets, and air brick into one simple unit. The unit fits over the window before it is fixed, and the difficulties of plugging and screwing to concrete or hard plaster, of fixing brackets, and of obtaining a neat joint between the pelmet board and plaster, are all eliminated. Designed by Williams & Williams Ltd., in collaboration with Frederick Gibberd, F.R.I.B.A.

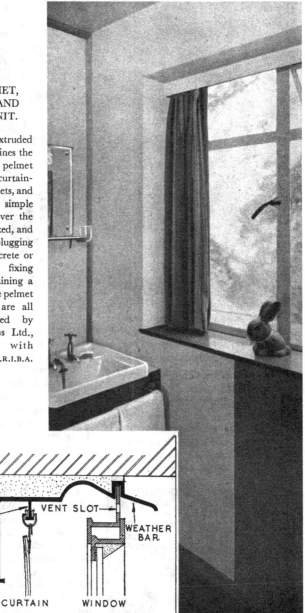

CURTAIN TRACK

VENT SLOT

WEATHER BAR.

PELMET

CURTAIN

WINDOW

PLATE XII

LIVING ROOM STORAGE UNITS.

These units, each about 3′ 6″ high by 3′ 9″ wide by 1′ 6″ deep, are intended to provide in a variety of possible arrangements the working and storage space needed in living rooms.

Each unit contains two compartments of which the lower is for general storage and enclosed by doors sliding on quiet plastic tracks, while the upper is either similar or has a fall front so that, with the appropriate fittings, it can become a desk, or a cupboard for gramophone records or drinks. By modification a unit can be adapted to special purposes —in the photograph the furthest one houses a radio set and gramophone, the radio controls and speaker below being covered by standard sliding doors when not in use and the automatic record changer, which when working is interesting to watch but should be sealed in to reduce needle scratch, being covered by a polished plate glass slide. Designer: R. D. Russell.

A CHEST IN MAHOGANY AND CELLULOSED WHITE WOOD. This is a small chest to stand against a wall or at the end of a bed. It is one of a series of articles of furniture designed for production in quantities. Certain dimensions were standardised for all the furniture so that the variety and sizes of drawers, trays and doors were reduced to a minimum, thereby simplifying the manufacturing process. Designer: Frederick Gibberd, F.R.I.B.A.

PLATE XIII

Top Right: Murphy Radio A.50 table set, cabinet designed by R. D. Russell. This set was one of the most expensive in the Murphy Radio range of its year (1938) so that the limitations in structure and material did not apply to the same extent as in the B.31 set (shown below). Honduras mahogany with a small flecked figure is used on the top and front. The ends of the cabinet are of straight grained Honduras mahogany veneered on plywood and the whole is finished with a deep eggshell polish, the wood being left in its natural colour.

Besides openings for speaker and scale, four controls and a tuning indicator had to be incorporated in the front. The apparent number of controls is reduced by combining them into two concentric pairs on either side of the scale, but even so the remaining openings, if cut in a plywood front and each framed by a separate escutcheon, would have produced a restless appearance. So a control panel was devised in glass with dark gunmetal coloured mirror plating on the back polished clear over scale and tuning indicator and drilled for the controls. The speaker fabric purposely woven in the same dark gunmetal colour with a thin white stripe is mounted on a frame which butts up to the control panel and continues right across to the other end of the cabinet. A narrow strip of white rubber cushions the control and speaker panels top and bottom.

In this cabinet the desired effect was obtained by a considerable amount of co-operation with mechanical and electrical designers in the early stages.

more pleasant in appearance and less vulnerable in use than uncompromising square corners would have been. The edges of the openings cut in the front are carefully smoothed and filled so that the plywood laminations will not show, and are then finished in ivory enamel. The speaker fabric is a purposely designed material, pale venetian red in colour, and is an important part of the whole carefully considered colour scheme. The positions of controls in relation to the scale and of the scale in relation to the speaker, the arrangement of the whole group within the framework of the cabinet and the proportions of the complete "works" are all subject to mechanical and electrical limitations and, to give a satisfactory result, must be influenced by the designer of the cabinet in close co-operation with the mechanical and electrical designers from the beginning. The design of the scale and control knobs is also of considerable importance to the cabinet designer and comes within his field.

(Reproduced by courtesy of Murphy Radio Ltd.)

Below: This set (also designed by R. D. Russell) was one of the lowest in price in the Murphy Range of its year (1937) and the cabinet, to keep in step, had to be simply made of inexpensive materials. The top and front are formed by bending one piece of plywood and, being the parts most in evidence, are veneered with well marked straight grained walnut left in its natural colour while the ends are of unveneered birch ply stained to a dark warm brown. By taking the plywood front right down to the table and splaying it up at each end finger-grips for lifting are provided, the expense of a visible and finished plinth is saved and the result is

PLATE XIV *Above:* A decanter and set of glasses made of flint glass. Designer: J. Hogan, R.D.I.

(Reproduced by courtesy of James Powell & Sons Ltd.)

Above, right: Individual small casseroles of cooking glass.

Below: A large casserole of cooking glass.
Designed by Harold Stabler, R.D.I., and Elizabeth Craig.

(Reproduced by courtesy of Phœnix Glass Ltd. Photographers, E. S. & A. Robinson, Bristol.)

PLATE XV *Above:* An admirable example of decorative industrial art: a jug and mug in the Wedgwood "Garden Implement" lemonade set. Designed by the late Eric Ravilious, A.R.C.A., N.R.D., for Josiah Wedgwood & Co., Ltd.

Below: A coffee set in matt white moonstone glaze, designed by Keith Murray, F.R.I.B.A., R.D.I., for Josiah Wedgwood & Co., Ltd.

PLATE XVI

Above: A bracket fitting specially designed for over-bed lighting in hospitals and nursing homes. The glass screws directly into the chromium plated metal base.

Left: A ceiling fitting consisting of a chromium plated base with a screw thread and opal glass which is threaded to fit into the metal base. Designed by A. B. Read, A.R.C.A., R.D.I., for Troughton & Young, Ltd.

Printed in the United States
by Baker & Taylor Publisher Services